Lecture Notes in Statistics

Vol. 1: R. A. Fisher: An Appreciation. Edited by S. E. Fienberg and D. V. Hinkley. xi, 208 pages, 1980.

Vol. 2: Mathematical Statistics and Probability Theory. Proceedings 1978. Edited by W. Klonecki, A. Kozek, and J. Rosiński. xxiv, 373 pages, 1980.

Vol. 3: B. D. Spencer, Benefit-Cost Analysis of Data Used to Allocate Funds. viii, 296 pages, 1980.

Vol. 4: E. A. van Doorn, Stochastic Monotonicity and Queueing Applications of Birth-Death Processes. vi, 118 pages, 1981.

Vol. 5: T. Rolski, Stationary Random Processes Associated with Point Processes. vi, 139 pages, 1981.

Vol. 6: S. S. Gupta and D.-Y. Huang, Multiple Statistical Decision Theory: Recent Developments. viii, 104 pages, 1981.

Vol. 7: M. Akahira and K. Takeuchi, Asymptotic Efficiency of Statistical Estimators. viii, 242 pages, 1981.

Vol. 8: The First Pannonian Symposium on Mathematical Statistics. Edited by P. Révész, L. Schmetterer, and V. M. Zolotarev. vi, 308 pages, 1981.

Vol. 9: B. Jorgensen, Statistical Properties of the Generalized Inverse Gaussian Distribution. vi, 188 pages, 1981.

Vol. 10: A. A. McIntosh, Fitting Linear Models: An Application of Conjugate Gradient Algorithms. vi, 200 pages, 1982.

Vol. 11: D. F. Nicholls and B. G. Quinn, Random Coefficient Autoregressive Models: An Introduction. v, 154 pages, 1982.

Vol. 12: M. Jacobsen, Statistical Analysis of Counting Processes. vii, 226 pages, 1982.

Vol. 13: J. Pfanzagl (with the assistance of W. Wefelmeyer), Contributions to a General Asymptotic Statistical Theory. vii, 315 pages, 1982.

Vol. 14: GLIM 82: Proceedings of the International Conference on Generalised Linear Models. Edited by R. Gilchrist. v, 188 pages, 1982.

Vol. 15: K. R. W. Brewer and M. Hanif, Sampling with Unequal Probabilities. vii, 164 pages, 1983.

Vol. 16: Specifying Statistical Models: From Parametric to Non-Parametric, Using Bayesian or Non-Bayesian Approaches. Edited by J. P. Florens, M. Mouchart, J. P. Raoult, L. Simar, and A. F. M. Smith. xi, 204 pages, 1983.

Lecture Notes in Statistics

Edited by D. Brillinger, S. Fienberg, J. Gani, J. Hartigan, and K. Krickeberg

16

Specifying Statistical Models

From Parametric to Non-Parametric, Using Bayesian or Non-Bayesian Approaches

Edited by J.P. Florens, M. Mouchart, J.P. Raoult, L. Simar, and A.F.M. Smith

---

Springer Science+Business Media, LLC

J.P. Florens
Université d'Aix-Marseille II
France

M. Mouchart
C.O.R.E.
Université Catholique de Louvain
Belgium

J.P. Raoult
Université de Rouen
France

L. Simar
Facultés Universitaires Saint-Louis
Bruxelles
Belgium

A.F.M. Smith
University of Nottingham
United Kingdom

AMS Classifications: 62-06, 62CXX, 62D05

Library of Congress Cataloging in Publication Data
Franco-Belgian Meeting of Statisticians (2nd: 1981:
Louvain-la-Neuve, Belgium)
Specifying statistical models (from parametric to
non-parametric, using Bayesian or non-Bayesian approaches)
(Lecture notes in statistics; 16)
1. Mathematical statistics—Congresses. I. Florens,
J.P. II. Title. III. Series: Lecture notes in statistics
(Springer-Verlag); 16.
QA276.A1F73 1981 519.5 82-19314

Proceedings of the Second Franco-Belgian Meeting of Statisticians held in
Louvain-la-Neuve (Belgium) on October 15–16, 1981.

With 24 Illustrations

9 8 7 6 5 4 3 2 1

ISBN 978-0-387-90809-0 ISBN 978-1-4612-5503-1 (eBook)
DOI 10.1007/978-1-4612-5503-1

CONTENTS

PREFACE

During the last decades, the evolution of theoretical statistics has been marked by a considerable expansion of the number of mathematically and computationaly tractable models. Faced with this inflation, applied statisticians feel more and more uncomfortable : they are often hesitant about their traditional (typically parametric) assumptions, such as normal and i.i.d., ARMA forms for time-series, etc., but are at the same time afraid of venturing into the jungle of less familiar models. The problem of the justification for taking up one model rather than another one is thus a crucial one, and can take different forms.

(a) Specification : Do observations suggest the use of a different model from the one initially proposed (e.g. one which takes account of outliers), or do they render plausible a choice from among different proposed models (e.g. fixing or not the value of a certain parameter) ?

(b) Model Approximation : How is it possible to compute a "distance" between a given model and a less (or more) sophisticated one, and what is the technical meaning of such a "distance" ?

(c) Robustness : To what extent do the qualities of a procedure, well adapted to a "small" model, deteriorate when this model is replaced by a more general one ? This question can be considered not only, as usual, in a parametric framework (contamination) or in the extension from parametric to non parametric models but also, within a non parametric framework, for evaluating the "weight" of some technical hypothesis (such as markovicity, knowledge of the stationary measure of a process, properties of a regression function, etc.).

(d) Adaptivity : Once one decides to step outside a traditional framework, is it possible to adapt procedures in current use within such a framework so as to obtain tractable procedures (e.g. the choice of a tractable prior measure in a non parametric framework, the computation of an approximate bayesian estimator instead of an intractable exact one, the enlarging of the null hypothesis w.r.t. which a statistic is parameter free, etc.) ?

The 12 lectures collected in these Proceedings were all presented at the "Rencontre Franco-Belge de Statisticiens" to an audience of research workers in applied and theoretical statistics. Different classifications could be possible for these papers : according to the degree of generality of their subjects (papers [1] to [5] were specially considered as methodology papers, even if they contain some specific results), according to the type of statistical theory to which they refer (papers [1], [2], [6], [7] and [8] belong to Bayesian statistics), according to their probabilistic framework (papers [4], [5], [11] and, partially, [2] and [9] are devoted to statistics on random processes; paper [7] uses the classical tools of linear statistics).

The papers are printed in the order of presentation at the conference. Papers [1] to [5] were presented the first day, oriented toward a larger audience. Some of the other papers are more technical. In this introduction, we have classified then according to the four types of methodological considerations that are listed above.

(a) Specification

Bayesian Statistics provide a natural framework for specification problems (and Box and Tiao's work in 1973 is quite enlightning in this respect). Roughly speaking, a "huge" model can always be considered as an union of smaller ones, and supplied with a prior measure. The consideration of the a posteriori measure over different small models does not present any theoretical difficulty; however, such unions of models may present some intricacies of interpretation (e.g. is there a common meaning for parameters having the same name inside the different models ?) and some computational difficulties which are analyzed in [1] (L. Simar); a review of some classical examples in this domain is given in [2] (A.F.M. Smith), where new results are presented for the treatment of outliers (in the univariate case, the different small models among which one has to choose are characterized by the number of possible lower and upper outliers; a multivariate generalization is sketched).

(b) Model Approximation

In Bayesian statistics, there are different ways of defining a distance between a model and a submodel; if the submodel is obtained by reduction on one only of the two component spaces (parameter space and sample space) one gets a notion of approximate sufficiency; various such notions are studied in [6] (J.P. Florens); if the reduction is made on the parameter space, ε-sufficiency (with ε sufficiently small) can be a justification for adopting the reduced model instead of the complete one.

In a non-bayesian framework, one can extend a given model by supposing that the parameter θ (initially considered as deterministic) may be submitted to a random pertubation in the neighbourhood of a fixed (but unknown) value; for i.i.d. observations (in the initial model), A. Hillion computes in [10] how the distance (variation distance, or Hellinger distance) between probabilities (for an infinite sequence of observations) in the initial model or in the perturbed model are bounded according to the magnitude of the perturbation; convergence results follow.

(c) Robustness

Classical statistical analyses of stationary time series rely heavily on the Box Jenkins approach for ARMA models; the specification of the order of these models is often heuristically and technically very difficult; in such cases, it is natural to propose non parametric estimators, for instance for the conditional expectation of $g(x_{n+s})$, given $(x_1, \ldots, x_n)$; convergence properties of such kernel estimators (based on x_n and on all couples $(x_i, g(x_{i+s}))$ $(1 \leqslant i \leqslant n-s)$ are studied in [5] (D. Bosq). Although heuristically markovian, these estimators rely only on mixing properties of the process and on hypothesis connecting the kernel and the stationary measure; in case this stationary measure is unknown, robustness properties (w.r.t. to bad choices of the kernel) are given in terms of the upper bound of the mean quadratic risk.

The same problem, for multivariate mixing processes, is studied in [11] (G. Collomb). In this paper, the estimators, called predictograms, are not based on general kernels, but on partitions of the space of observations $(x_1, \ldots, x_n)$ and make use of the k-uple $(x_{n-k+1}, \ldots, x_n)$ and of all the (k+1)-uples $(x_{i-k+1}, \ldots, x_i, g(x_{i+s}))$ $(k \leq i \leq n-s)$. Comparison of the speeds of convergence given in [5] and [11] can be useful for the choice of the number of successive observations used in the computations (i.e. typically, though the results do not rely directly on Markov properties, the order of markovicity assumed in the model).

In [4] (P. Doukhan), attention is focused on autoregressive Markov processes $(x_{n+1} = f(x_n) + \varepsilon_n)$; consistent kernel estimators of f are known, under hypothesis of regularity on f $(\|f\|_\infty + \|f'\|_\infty < \infty)$; their mean quadratic risk is of order $n^{-2/3}$, up to multiplicative constants in which both f and the variance σ^2 of the noise ε_n intervene; simulations are made, in [4], in order to study the deterioration of the qualities of these estimators for some "bad" functions of regression (i.e. not satisfying the hypothesis of the convergence theorem) and their sensitivity to the variations of σ^2 and to the shape of f (e.g. a periodic f, with its period small w.r.t. σ, constitutes a model which is indistinguishable from an i.i.d. model).

In the i.i.d. case, it is a classical problem to test (by means of "least favourable couples") a ball $B(P,r)$ against an other $B(P',r')$ (the radii being measured with the Hellinger distance). The error probabilities are known to decrease exponentially to 0 ; in [9] (L. Birge) explicit computations of the way these error probabilities increase when the "reality" (not necessarily i.d.) lies outside of one of these balls are given; for example, among n observations, all laws slightly outside $B(P,r)$, or a small number among the n observations distinctly outside $B(P,r)$. As a consequence of these computations, one gets properties of robustness w.r.t. the choice of radii; in other words one gets an answer to the question : what is the penalty if an attempt of robustification, leading to a model with the ball $B(P,r)$ instead of its center P, has not been strong enough ?

(d) Adaptivity

The classical robust tests between neighbourhoods of given probabilities, as seen in the last paragraph of (c) above, are no more tractable when the centers of these neighbourhoods do not correspond to i.i.d. models. In such cases (i. non i.d., or Markov) one has to use generalizations of the Hellinger distance; balls are not convex any more, and the "least favorable couple" argument does not hold. In the same paper [9], L. Birge provides adaptations of the classical tests of neighbourhoods, keeping the fundamental property of exponentially decreasing error probabilities.

In multivariate non parametric testing of hypothesis, it is sometimes proposed that test functions be constructed by means of parameter free statistics whose law is multidimensional uniform with respect to any probability in the null hypothesis (this is for instance quite usual for testing multidimensional normality). Such parameter-free statistics are constructed by using sufficient statistics whose laws are absolutely continuous. In [3] (J.P. Raoult, D. Criticou, D. Terzakis) this type of method is adapted to the non absolutely continuous case (tests are then randomized tests).

Bayesian non parametric theory, which had been neglected for a long time, became tractable essentially after the introduction (Ferguson, Doksum, 1973) of Dirichlet processes which allow easy computations of posterior probabilities on the set of all distribution functions on $\mathbb{R}$. In [8] (J.M. Rolin) attention is focused on the "general theory of processes" properties which make these computations feasible. This paper characterizes the Dirichlet process by stochastic independence properties rather than by distributional properties. Some extensions of these independence properties lead to consider neutral processes as a natural class of tractable prior probabilities.

A final adaptivity problem, in a Bayesian framework, is the one studied in [7] (M. Mouchart et L. Simar). Computation of the posterior expectation of the parameter, $E(\theta/x)$ is often difficult once one decides to step outside a traditional model. Least squares approximations within more general models are shown to provide tractable adaptation; the role of exchangeability conditions is studied in this respect.

Finally, paper [12] (J. Benasseni) is devoted to principal component analysis : once chosen weights for the observations (in order to define a distance between the variables), the usual procedures make use of the centering of the observations which is connected to those weights but, this connection does not exist any more if the computations are done by using weights and centering constant which are respectively the coefficients appearing in robust estimates of mean and variance. It is shown that however this procedure can still be considered as principal component analysis provided the distance between the variable is suitably defined.

The editors wish to thank the Fonds National de la Recherche Scientifique, the Center for Operations Research and Econometrics (C.O.R.E. - Université Catholique de Louvain) and the Séminaire de Mathématiques Appliquées aux Sciences Humaines (SMASH - Facultés Universitaires Saint-Louis - Bruxelles) for their support to this meeting.

The Editors,

Florens, J.-P.
Mouchart, M.
Raoult, J.-P.
Simar, L.
Smith, A.F.M.

LIST OF PARTICIPANTS

BALLET, M.F.
Boels et Bégault
Rue des Blancs Chevaux 7-301
1348 Louvain-la-Neuve

BENASSENI, J.
CRIG Montpellier
"Résidence les Rainettes" Bat. C
Rue de Las Sorbes, 1250
Montpellier - France

BENMANSOUR, D.
Université de Rouen
Cité Panorama, 112 Flaubert
76130 Mont-Saint-Signon

BERTRAND, D.
U.C.L.
Chemin Long, 213
1310 La Hulpe

BIRGE, L.
Université de Paris X, Nanterre
Boulevard Magenta, 93
75010 Paris - France

BIRONT, E.
U.C.L.
Chaussée de Boondael, 306
1050 Bruxelles

BOSQ, D.
Université de Lille I
Rue Dunois, 74
75013 Paris - France

CARLETTI, G.
IRSIA - Bureau de Biométrie
Rue de Roncia, 23
5800 Gembloux

COCCHI, D.
CORE - U.C.L.
Rue de la Neuville, 30
1348 Louvain-la-Neuve

COLLOMB, G.
U.P.S.
Rue de l'Homme Armé, 1
31000 Toulouse - France

DEHANDTSCHUTTER, M.
U.C.L.
Rue des Blancs Chevaux, 1/101
1348 Louvain-la-Neuve

de LEVAL, N.
Fac. de Médecine - U.C.L.
Rue des Cours, 13
5865 Walhain

DEPRINS, J.
U.L.B.
Stanley, 7
1980 Tervuren

DEPRINS-VANHECKE, D.
Facultés Universitaires St Louis
Avenue des Prisonniers de Guerre, 5
1490 Court-Saint-Etienne

DOUKHAN, P.
Université de Rouen
Rue de la Croix Nivert, 214
75015 Paris - France

DREZE, J.
CORE - U.C.L.
Voie du Roman Pays, 34
1348 Louvain-la-Neuve

EMBRECHTS P.
K.U.L.
Fonteinstraat, 58/2
3031 Oud-Heverlee

FLORENS, J.-P.
Université d'Aix Marseille
Le Corbusier, 229 -Bd Michelet
13008 Marseille - France

FOURDRINIER, D.
Université de Rouen
Rue de Beauvoisine, 12
76000 Rouen - France

GOVAERTS, B.
CORE - U.C.L.
Avenue M. César, 89
1970 Wezembeek

GRANCHER, G.
Rue Roger Gobbé, 354
76230 Bois-Guillaume - France

GUERIT, Y.
Crédit Communal de Belgique
Rue du Grand Marais, 11
7410 Mons

HALLIN, M.
U.L.B.
Avenue de Catus, 6
1310 La Hulpe

HILLION, A.
Université de Bretagne Occidentale
Rue Joseph le Brix, 20
29200 Brest - France

HOUTMAN, A.
CORE - U.C.L.
Avenue Chataigniers, 3
1640 Rhode-Saint-Genèse

JANSSEN, J.
U.L.B.
Avenue J. Stobbaerts, 59
1030 Bruxelles

JOYEUX, R.
CORE - U.C.L. et Cornell Univ. (U.S.A.)
Rue des Pêcheries, 103 Boîte 50
1160 Bruxelles

KESTEMONT, M.-P.
F.U.S.L. et U.C.L.
Rue Klakkedelle, 68
1200 Bruxelles

KUPPER, J.-M.
Grand Monchaut, 74
7890 Ellezelles

LUBRANO, M.
CORE - U.C.L.
Rue du Sablon, 32/405
1348 Louvain-la-Neuve

MACHTELINCKX
U.L.B.
Avenue A. Giraud, 48
1030 Bruxelles

MAES, J.-M.
Avenue de Tervuren, 441
1150 Bruxelles

MALBECQ, W.
U.L.B.
Avenue Général Lebon, 113, Bte 7
1160 Bruxelles

MERCIER , F.
U.C.L.
Rue A. Delvaux, 17
6040 Charleroi

MIKHAEL, A.
Université de Rouen
Fac. des Sciences
Bâtiment de Mathématiques
76130 Mont-Saint-Aignan - France

MOUCHART, M.
CORE - U.C.L.
Rue Haute, 47
1348 Louvain-la-Neuve

ORSI, R.
CORE - U.C.L.
Rue de la Neuville, 30
1348 Louvain-la-Neuve

PARIS, J.
U.C.L.
Boulevard Paul Janson, 88
6000 Charleroi

RAOULT, J.-P.
Université de Rouen
Avenue Ganbetta, 15
92410 Ville d'Avray - France

REY, W.
M.B.L.E. - Bruxelles
Boulevard du Souverain, 138, Bte 19
1190 Bruxelles

ROLIN, J.-M.
U.C.L.
Rue Elizabeth, 1
5800 Gembloux

RUTGEERTS, A.-M.
ADPU - U.C.L.
Rue des Blancs Chevaux, 1/401
1348 Louvain-la-Neuve

SIMAR, L.
Facultés Universitaires St Louis
Avenue des Chênes, 16
5870 Mont-Saint-Guibert

SMITH, A.F.M.
University of Nottingham
Nether St., 41
Bekston, Nottingham, U.K.

SPIES, J.-M.
S.N.C.B.
Clos des Pommiers, 29
1310 La Hulpe

VAES, T.
IAG - U.C.L.
Rue de Parme, 45
1060 Bruxelles

VAN CUTSEM, B.
Univ. Scient. et Méd. de Grenoble
Avenue Louis Noiray Corenc, 7.
38700 La Tronche - France

VANDAUDENARD, R.
U.C.L.
Rue Brichaut, 16
1030 Bruxelles

PROTECTING AGAINST GROSS ERRORS : THE AID OF BAYESIAN METHODS

by

Léopold Simar

Facultés Universitaires Saint-Louis, Bruxelles
and Center for Operations Research and Econometrics
Université Catholique de Louvain

Abstract

A statistical model is characterized by a family of probability distribution functions. All inferences are then conditional on the hypothesis formalised by this family.

The statistician often needs to protect himself against the consequences of a gross error relative to the basic hypothesis : either a specification error for the functionnal form of $p(x|\theta)$, or the treatment of outliers. It will be shown in this paper that the Bayesian approach offers a natural framework for treating this kind of problem. Different methods are presented : robustness analysis considering the sensitivity of inference to the model specification; and approximations to Bayesian solutions which are for a large class of models and sometimes preferable to the exact solutions valid only for a particular model.

Key-words : Bayesian sensitivity, Robustness analysis.

AMS/MOS : Primary 62A15, Secondary 62G35.

Acknowledgements

I would like to express my thanks to M. Mouchart with whom I have had numerous discussions on the subject of this paper and to J.-M. Rolin for his help in revising the paper. Errors and shortcomings are exclusively my responsibility.

1. INTRODUCTION

Many statistical procedures are based on statistical models which specify under which conditions the data are generated. Usually the assumption is made that the set of observations $x_1, \ldots, x_n$ is a set of (i) independent random variables (ii) identically distributed with common p.d.f. $p(x_i|\theta)$. Once this model is specified, the statistician tries to find optimal solutions to his problem (usually related to the inference on a set of parameters $\theta \in \Theta \subset \mathbb{R}^k$, characterizing the uncertainty about the model).

The procedure just described is not always easy to carry out. In fact when confronted with a set of data three attitudes are possible :

(1) The statistician may be a "pessimist" who does not belief in any particular model $p(x|\theta)$. In this case he must be satisfied with descriptive methods (like exploratory data analysis) without the possibility of inductive inference.

(2) The statistician may be an "optimist" who strongly believes in one model. In this case the analysis is straightforward and optimal solutions may often be easily obtained.

(3) The statistician may be "realist" : he would like to specify a particular model $p(x|\theta)$ in order to get operational results but he may have either some doubt about the validity of this hypothesis or some difficulty in choosing a particular parametric family.

Let us illustrate this kind of preoccupation with an example. Suppose that the parameter of interest is the "centre" of some population. In many situations, the statistician may argue that, due to a central limit effect, the data are generated by a normal p.d.f. In this case the problem is restricted to the problem of inference on μ, the mean of the population. But in some cases, he may have some doubt about these central limit effects and may suspect some skewness and/or some kurtosis or he may suspect that some observations are generated by other models (leading to the presence of outliers).

In this context three types of question may be raised in order to avoid gross errors in the prediction, or in the inference :

(i) Does the optimal solution, computed for an assumed model $p(x|\theta)$, still have "good" properties if the true model is a little different ?

(ii) Are the optimal solutions computed for other models near to the original one really substantially different ?

(iii) Is it possible to compute (exactly or approximatively) optimal solutions for a wider class of models based on very few assumptions ?

The first question is concerned with the sensitivity of a given criterion to the hypothesis (criterion robustness). In the second question, it is the sensitivity of the inference which is analysed (inference robustness). The last question may be viewed as a tentative first step towards the development of nonparametric methods (i.e. methods based on a very large parametric space).

The object of the paper is to provide some insight into the problems raised by the two latter questions. It will be shown that a sensitivity (robustness) analysis is natural in a Bayesian framework and that distribution-free methods (valid for a large class of models) can be obtained. Different approaches will be considered, which can be schematically presented as follows.

Let $\mathcal{F}$ be the class of all distribution functions on $\mathbb{R}$ and let $\mathcal{F}_i$, $i = 0,1,2...$ be a family of parametric classes of distributions on $\mathbb{R}$.

(1) Analysis in the neighborhood of a given model $\mathcal{F}_0$

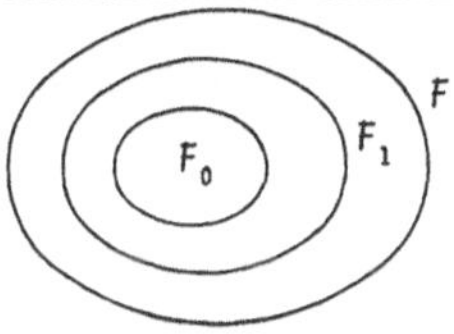

Let $\mathcal{F}_1$ be a wider class of models including $\mathcal{F}_0$. Working with $\mathcal{F}_1$ we obtain more robust methods and a sensitivity analysis will be considered in this neighborhood of $\mathcal{F}_0$.

(2) Mixtures of models

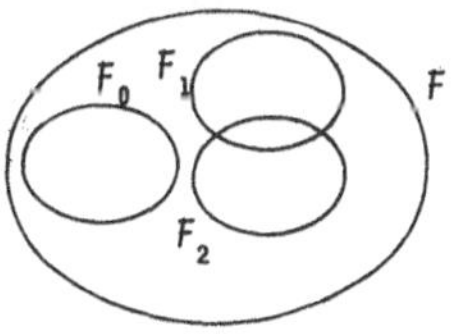

Working with $\mathcal{F}_0 \cup \mathcal{F}_1 \cup \mathcal{F}_2$ ..., allows us to obtain even more robust methods and provides a framework for analysing the sensitivity of the inference to the choice of a model.

(3) Approaches to nonparametric methods

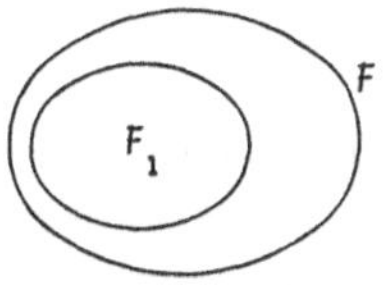

Let $\mathcal{F}_1$ be a "large" class of models so that "almost distribution-free" methods are obtained.

It will be shown that Bayesian methods provide a natural framework for posing the problem but that two kinds of difficulties may be encountered : computational problems and problems of interpretation (choice of parametrisation; assessment of prior distribution, etc.).

2. INFERENCE ROBUSTNESS

In this section, it will be seen that the Bayesian formulation of the approaches (1) and (2) above, implies the introduction of some hyperparameters either describing the neighborhood of a given model, or tracing a family of different model. Therefore, the mathematical formulation and the difficulties are essentially the same. However, the structure of the priors will be quite different.

2.1. Neighborhood of a model

The idea is to consider the data to be generated by some member of a wider class of distribution functions F_1, including as a subset F_0 the family of distribution functions represented by the original model. The date density of F_1 will now be written $p(x|\theta,\nu)$ where θ is the parameter of interest and ν is a new parameter characterising some neighborhood of the basic model; if ν_0 is the value of ν corresponding to the basic model, then $\forall\, \theta \in \Theta$, $p(x|\theta,\nu_0)$ is a member of F_0.

Once the prior on the new parameter space $p(\theta,\nu)$ is specified, the Bayesian analysis is straightforward :

$$p(\theta,\nu|x) \propto p(x|\theta,\nu)\; p(\theta,\nu). \tag{2.1}$$

The following outputs of the analysis are of interest :

(i) $p(\theta|x)$ represents the overall (marginal) information on θ. So the inference on θ is now based on a wider class of models and this provides more "robust" procedures.

(ii) $p(\nu|x)$ allows one to analyse the plausibility of several models including the simpler model $(\nu = \nu_0)$ and it shows how the uncertainty about the model, represented by $p(\nu)$, is transformed by the data.

(iii) $p(\theta|x,\nu)$ computed for different value of ν, reveals the sensitivity of the inference about θ to small departures from $\nu = \nu_0$.

Thus, it appears that the Bayesian approach provides (at least from a theoretical point of view) the natural framework to think about robustness problems. However, there may be some practical difficulties in performing this analysis. These difficulties will be discussed after the two following examples.

Example 1 (Box-Tiao (1973)) : The family of exponential power densities

The data density (for one observation) is written as :

$$p(x|\mu,\sigma,\nu) = \frac{[\Gamma(3/2(1+\nu))]^{1/2}}{(\nu+1)\;[\Gamma(1/2(1+\nu))]^{3/2}}\,\sigma^{-1}\exp\left\{-\left[\frac{\Gamma(3/2(1+\nu))}{\Gamma(1/2(1+\nu))}\right]^{1/1+\nu}\left|\frac{x-\mu}{\sigma}\right|^{2/1+\nu}\right\} \tag{2.2}$$

where $\mu \in \mathbb{R}$, $\sigma \in \mathbb{R}^+$, $-1 < \nu \leq +1$. This may be written as

$$p(x|\mu,\sigma,\nu) \propto \sigma^{-1} \exp\left\{- a(\nu) \left|\frac{x-\mu}{\sigma}\right|^{2/1+\nu}\right\} \tag{2.3}$$

and represents a class of symmetric distributions including the normal $(\nu = 0)$, the double exponential $(\nu = 1)$ and the uniform (when ν tends to -1). The different posterior p.d.f. are obtained by numerical integration and Box and Tiao (1973) propose a particular form for $p(\nu)$ which facilitates these computations.

Example 2 (Relles-Rogers (1977)) : The family of student densities

$$p(x|\mu,\sigma,\nu) = \frac{\nu^{1/2}}{B\,(1/2,1/2\nu)}\, \sigma^{-1} \left[1 + \nu\left(\frac{x-\mu}{\sigma}\right)^2\right]^{-(1+\nu)/2\nu} \tag{2.4}$$

where $\mu \in \mathbb{R}$, $\sigma \in \mathbb{R}^+$, $0 \leq \nu \leq 1$. In this parametrisation, ν is the inverse of the degrees of freedom; when $\nu = 0$ we obtain the normal density and for $\nu = 1$, the Cauchy density. Relles and Rogers (1977) analyse by simulation the robustness properties of the estimator of μ obtained from $p(\mu|x)$ versus some classical robust estimators for location parameters. They pointed out the good performances of the Bayesian estimator.

Comments

The analysis presented above seems elegant and very attractive. But, in addition to the computional difficulties, there may be a lot of problems in assessing the prior p.d.f. $p(\theta,\nu)$. Usually, this prior is specified through the following decomposition :

$$p(\theta,\nu) = p(\theta|\nu)\, p(\nu) \tag{2.5}$$

and the questions are : how to assess $p(\theta|\nu)$ for different values of ν in a coherent way ? Does θ represent the same physical characteristic of the population for different value of ν ? It is reasonable to let θ and ν be prior independent ? In the latter case $p(\theta|\nu) = p(\theta)$ and the specification of $p(\theta,\nu)$ would certainly be facilitated. Thus it appears that the choice of the parametrisation seems to be crucial.

In example 1 (exponential power family) μ and σ represent the mean and standard deviation of the population for each value of ν and Box-Tiao (1973) suppose that (μ,σ) and ν are a priori independent. The following question may be raised : is it reasonable that the prior on σ is the same for $\nu = 0$ (the normal case) and for ν approaching -1 (the uniform case) ?

In example 2 (student family), the choice of the parametrisation is important. The student family could have been parametrised as follows :

$$p(x|\mu,\alpha,\nu) = \frac{1}{B\,(1/2,1/2\nu)}\, \alpha^{-1} \left[1 + \left(\frac{x-\mu}{\alpha}\right)^2\right]^{-1/2\nu - 1/2} \tag{2.6}$$

when $\alpha \in \mathbb{R}^+$ is a different choice of scale parameter.

We have the following relations between α and σ :

$$\mathrm{Var}\,(x|\mu,\sigma,\nu) = \frac{\sigma^2}{1-2\nu} \quad \text{if} \quad \nu < 1/2; \tag{2.7}$$

$$\mathrm{Var}\,(x|\mu,\alpha,\nu) = \frac{\nu\alpha^2}{1-2\nu} \quad \text{if} \quad \nu < 1/2. \tag{2.8}$$

The question may be raised : is it more reasonable to have a priori independence between σ and ν or between α and ν ? (Relles-Rogers (1977) have chosen $p(\sigma,\nu) = p(\sigma)$). One could have chosen $\gamma = \sigma^2/1-2\nu$ as parameter (with $\nu < 1/2$). In this case μ and γ would have the same interpretation for all ν but the problems of independence are still present. Note that Box and Tiao (1973) have pointed out that in the location scale family, a scale parameter is arbitrarily defined up to a multiplicative constant. If $\alpha = f(\nu)\,\sigma$, the prior independence between σ and ν is lost for α and ν.

More generally, it seems reasonable that prior opinion about θ must influence one's opinion about the possible models. This appears when writing $p(\nu|x)$ even if $p(\theta,\nu) = p(\theta)\,p(\nu)$, since

$$p(\nu|x) \propto p(\nu) \int_\Theta p(x|\theta,\nu)\,p(\theta)\,d\theta. \tag{2.9}$$

No general answers can be given to this kind of question, but the comments above show that the elegance of the theoretical development should not hide the practical problems of interpretation.

2.2. Mixtures of models

Another way of enlarging the class of models considered is to extend the ideas of the preceding section in the following way; let the data density be written as follows

$$p(x|\theta_\lambda,\lambda) \tag{2.10}$$

where $\theta_\lambda \in \Theta_\lambda$, $\lambda \in \Lambda$, and where, for each $\lambda \in \Lambda$ (Λ is typically discrete), $p(x|\theta_\lambda,\lambda)$ is a particular parametric density with parameter space Θ_λ. Here again, in addition to the prior densities $p(\theta_\lambda|\lambda)$ a prior probability $p(\lambda)$ for each model must be specified. As before the Bayesian analysis is elegant and straightforward :

$$p(x|\lambda) = \int_{\Theta_\lambda} p(x|\theta_\lambda,\lambda)\,p(\theta_\lambda|\lambda)\,d\theta_\lambda. \tag{2.11}$$

We can obtain

$$p(\theta_\lambda|\lambda,x) = \frac{p(\theta_\lambda|\lambda)\,p(x|\theta_\lambda,\lambda)}{p(x|\lambda)}. \tag{2.12}$$

We also have the posterior probabilities for each model :

$$p(\lambda|x) = \frac{p(\lambda)\,p(x|\lambda)}{\sum_{\lambda\in\Lambda} p(\lambda)\,p(x|\lambda)}. \tag{2.13}$$

In this general presentation, it appears that the parameter space Θ has the particular structure

$$\Theta = \underset{\lambda \in \Lambda}{\vee} \Theta_\lambda = \{(\lambda, \theta_\lambda) \mid \lambda \in \Lambda,\ \theta_\lambda \in \Theta_\lambda\} \tag{2.14}$$

and therefore the problems of coherence in the specification of the prior distribution Θ are still more important than before.

If a simpler parametrisation can be chosen, such that $\theta_\lambda = \theta$ for all $\lambda \in \Lambda$, we have exactly the same formulation as in section 2.1. However if Λ is discrete, λ can no longer be interpreted as a parameter of perturbation. A sensitivity analysis is obtained by comparing $p(\theta|\lambda,x)$ for different values of λ; moreover, an inference based on $p(\theta|x)$ is certainly robust. But, even in this case, the problem of the interpretation of the common parameter θ under different models is still crucial. This approach has been used by Smith (1977) where $\theta = (\mu,\sigma)$ is a location-scale parameter for each model considered (the uniform, the normal, the double-exponential, etc.) and the questions are again : are μ and σ a priori independent of λ ? Is it reasonable to give to (μ,σ) the same physical meaning for different values of λ ?

The following example shows the sensitivity of inferences under different choices of the parametrisation.

Example

Suppose the experimenter is interested in θ the "centre" of the population, but he hesitates between a normal and a log-normal model.

$$p(x|\mu,\sigma,\lambda = 1) = \frac{1}{\sqrt{2\pi}\,\sigma} \exp\left\{-\frac{1}{2\sigma^2}(x-\mu)^2\right\}; \tag{2.15}$$

$$x \in \mathbb{R},\quad \mu \in \mathbb{R},\quad \sigma \in \mathbb{R}^+$$

and

$$p(x|\alpha,\beta,\lambda = 2) = \frac{1}{\sqrt{2\pi}\,\beta\, x} \exp\left\{-\frac{(\log x - \alpha)^2}{2\beta^2}\right\}; \tag{2.16}$$

$$x \in \mathbb{R}^+,\quad \alpha \in \mathbb{R},\quad \beta \in \mathbb{R}^+.$$

Since θ has a common physical meaning, it may be reasonable to assume prior independence between θ and λ.

$$p(\theta|\lambda = 1) = p(\theta|\lambda = 2) = p(\theta). \tag{2.17}$$

If $\lambda = 1$, θ is simultaneously the mean μ, the mode and the median of the population, so that $p(\mu)$ can be assigned unambigously. But if $\lambda = 2$ we have :

$$E(x|\alpha,\beta,\lambda = 2) = e^{\alpha+\beta^2/2} \tag{2.18}$$

$$\text{Mode } (x|\alpha,\beta,\lambda = 2) = e^{\alpha-\beta^2} \tag{2.19}$$

$$\text{Median } (x|\alpha,\beta,\lambda = 2) = e^{\alpha} \tag{2.20}$$

so that the prior $p(\theta)$ will induce different priors on (α,β) according to the choice of the statistical representation of θ under the log-normal model.

Here again no solutions are proposed but those problems must be borne in mind when using this approach.

The above model $p(x|\theta_\lambda,\lambda)$ is a particular case of a more general model where the sampling distribution is a mixture of the different models

$$p(x|\{\theta_\lambda\ \pi_\lambda : \lambda \in \Lambda\}) = \sum_{\lambda\in\Lambda} p(x|\theta_\lambda,\lambda)\ \pi_\lambda \tag{2.21}$$

where $\pi_\lambda \geqslant 0\ \forall\ \lambda \in \Lambda$

$$\sum_{\lambda\in\Lambda} \pi_\lambda = 1 \text{ a.s.}$$

In this case, in addition to the prior on θ_λ, $\lambda \in \Lambda$, a prior on the weights π_λ has to be specified (for example, a Dirichlet distribution). Here the difficulties are mainly of computational order since the likelihood of a sample of size n will be expressed as sum of $(\text{Card } \Lambda)^n$ terms. Smith and Makov (1978) propose approximate solutions for the estimation of π_λ which are easy to compute and are shown to have some convergence properties. This kind of model (2.21) can be used to treat the problem of outliers. For instance the contaminated normal

$$p(x|\mu,\sigma,\pi,k) = \pi\ \varphi(x|\mu,\sigma) + (1 - \pi)\ \varphi(x|\mu,k\sigma) \tag{2.22}$$

where $\pi \geqslant 0$, $k \geqslant 1$ and $\varphi(x\ .,.)$ is the normal p.d.f., has been used when the presence of outliers is suspected (Box-Tiao (1968) with known k, and more recently by Naylor (1982)).

3. "DISTRIBUTION-FREE" METHODS

In this section we will briefly mention two kinds of approach which allow one to develop distribution-free procedures in the sense that there is no need to specify completely the data generating process. In this respect the procedures are robust since they are valid for a large class of models.

3.1. Least squares approximations

Let $\theta \in R^k$ be the parameter of interest and $x \in R^n$ be the sample information. The least squares (L.S.) approximation of θ is the linear function of x which minimizes the expected quadratic error :

$$\min_{A,b}\ E_{\theta,x}\ [(\theta - (Ax + b))'\ (\theta - (Ax + b))] \tag{3.1}$$

where $A : k \times n$ and $b : k \times 1$.
The solution is known to be $\theta^*(x)$:

$$\theta^*(x) = E(\theta) + V_{\theta x} V_{xx}^{-1}[x - E(x)]. \tag{3.2}$$

The error of approximation $\eta = \theta - \theta^*(x)$ has the following properties

$$E(\eta) = 0 \tag{3.3}$$

$$V(\eta) = V_{\theta\theta} - V_{\theta x} V_{xx}^{-1} V_{x\theta} \tag{3.4}$$

where $V_{\theta\theta} = V(\theta)$
$V_{\theta x} = Cov(\theta, x')$
$V_{xx} = V(x)$.

The different moments appearing in (3.2) are usually computed through the following decomposition :

$$E(x) = E(E(x|\theta)) \tag{3.5}$$

$$V_{\theta x} = E(\theta\, E(x'|\theta)) - E(\theta)\, E(x') \tag{3.6}$$

$$V_{xx} = E(V(x|\theta)) + V(E(x|\theta)). \tag{3.7}$$

It must be noted that $\theta^*(x)$ is also the L.S. approximation of $E(\theta|x)$, the exact Bayesian solution w.r.t. quadratic loss.

The interest of the method is that for computing $\theta^*(x)$ and for evaluating the accuracy of the approximation via $V(\eta)$ there is no need to specify completely the joint distribution of (θ, x) but only the first two moments; this means that the procedure is not only robust w.r.t. the data generating process but also w.r.t. to the prior on θ. The idea is that L.S. acts as a smoothing procedure and that $\theta^*(x)$ is certainly less sensitive to variation of a given model than the exact solution.

Stone (1963) introduced this method in order to obtain robust Bayesian estimates for the mean of a process. It has been used by Goldstein (1975 a,b) as approximations of Bayesian solutions in nonparametric problems. A more systematic analysis with applications can be found in Mouchart-Simar (1980) and (1982).

3.2. Nonparametric approaches

A nonparametric model can be described as a model based on a parametric space so large that it cannot be characterized by a finite number of parameters. Typically the class of distribution functions considered is the class of all the distribution functions on the support of x (say $\mathbb{R}$)

$$\mathcal{F} = \{F \mid F \text{ is a distribution function on } \mathbb{R}\}. \tag{3.8}$$

The Bayesian analysis of such a model is rather complicated but several approaches have been proposed to handle random distribution functions; mainly : Dirichlet processes, Neutral processes and Tailfree processes (Ferguson (1973) and (1974) and Doksum (1974)). An elegant and simple presentation of neutral processes can be found in Rolin (1982). For further references see Simar (1982). In this section, we show how a very simple model allows one to get Bayesian solutions for a fairly large class of models which have good properties w.r.t. the larger model F given by (3.8). Further it introduces naturally the Dirichlet processes which are not discussed in this paper.

Suppose that the sample space $\mathbb{R}$ is partitionned into the intervals $A_1, \ldots, A_{k+1}$ and denote by Π_i the probability that an observation x falls in A_i :

$$\Pi_i = \text{Prob}\,(x \in A_i \mid F). \tag{3.9}$$

Consider now the following parametric space

$$\Pi = \{\Pi_1, \ldots, \Pi_{k+1} \mid \sum_{i=1}^{k+1} \Pi_i = 1, \;\; \Pi_i \geqslant 0\}. \tag{3.10}$$

There is not a one-to-one correspondence between F and Π but the analysis on Π is straightforward : it is the inference in a multinomial process where $n(A_i)$, the number of observations falling in A_i is a set of sufficient statistics, and the natural conjugate family on Π is the Dirichlet family of densities. So that, as far as the estimation of Π_i or the estimation of functions of Π_i and A_i is concerned, there is no problem in handling the inferences. For example if a center of the population is defined as

$$\mu(F) = \sum_{i=1}^{k+1} \ell(A_i)\, \Pi_i \tag{3.11}$$

where $\ell(A_i)$ denotes a centre of A_i, the Bayesian estimate of μ w.r.t. quadratic loss will be

$$\mu^*(F) = \sum_{i=1}^{k+1} \ell(A_i)\, E(\Pi_i \mid x_1, \ldots, x_n).$$

The solutions so obtained are in fact valid for all distributions $F \in F$ which are piecewise linear on $A_1, \ldots, A_{k+1}$ or more generally which give a weight Π_i on A_i. The problem is that the particular partition $A_1, \ldots, A_{k+1}$ is arbitrarily chosen and one would like to have coherent solutions for different choices of A_i. Two answers can be given to this kind of concern.

First it has been shown that under some conditions (on the loss function), the Bayesian solutions obtained w.r.t. the parameter space Π are *Mixed Bayes-Minimax* solutions (Doksum (1972)) to the general problem i.e. w.r.t. the parameter space F. In particular this ensures that when refining the partition $A_1, \ldots, A_{k+1}$, the solutions converge to the Bayesian solution one could obtain w.r.t. the parameter space F.

The second answer is the non-parametric approach to Bayesian statistics. The above coherence will be ensured if the distribution on $\Pi_1, \ldots, \Pi_{k+1}$ above for given $A_1, \ldots, A_{k+1}$ are in fact particular marginal distributions one could obtain from a random distribution function (defined on an adequate measurable space) with support F. For instance if the random distribution function F belongs to a Dirichlet process, Ferguson (1973) proved that the Kolmogoroff consistency conditions are satisfied. The main idea is that the parameters of the Dirichlet prior on $\Pi_1,\ldots,\Pi_{k+1}$ are chosen according to a finite non-negative measure ν on $(\mathbb{R},\beta)$ in the following way

$$\alpha_i = \frac{\nu(A_i)}{\nu(\mathbb{R})} \qquad i = 1, \ldots, k+1.$$

This roughly defines the Dirichlet process with parameter ν. For any partition $A_1, \ldots, A_{k+1}$, α_i is the prior expectation of Π_i; the posterior expectation of Π_i can then be written :

$$E(\Pi_i \mid x_1, \ldots, x_n) = p_n \frac{\nu(A_i)}{\nu(\mathbb{R})} + (1 - p_n) \frac{n(A_i)}{n}$$

where $p_n = \frac{\nu(\mathbb{R})}{\nu(\mathbb{R}) + n}$. In this notation it appears that $\nu(\mathbb{R})$ can be interpreted as the weight of the prior information while the shape of $\nu((-\infty,x])$ represents the prior beliefs on the shape of $F(x)$ i.e. the shape of the predictive of an observation x.

REFERENCES

[1] Box, G.E.P. and G.C. Tiao (1968), "A Bayesian Approach to Some Outlier Problems", *Biometrika 55*, 119.

[2] Box, G.E.P. and G.C. Tiao (1973), *Bayesian Inference in Statistical Analysis*, Addison-Welsey.

[3] Doksum, K.A. (1972), "Decision Theory for Some Nonparametric Models", *Proceedings Sixth Berkeley Symposium on Mathematical Statistics and Probability 1*, 331-341.

[4] Doksum, K.A. (1974), "Tailfree and Neutral Random Probabilities and their Posterior Distributions", *Ann. Probab. 2*(2), 183-201.

[5] Ferguson, T.S. (1973), "A Bayesian Analysis of Some Nonparametric Problems", *Ann. Stat. 1*, 209-230.

[6] Ferguson, T.S. (1974), "Prior Distribution on Spaces of Probability Measures", *Ann. Stat. 2*, 615-629.

[7] Goldstein, M. (1975a), "Approximate Bayes Solutions to Some Nonparametric Problems", *Ann. Stat. 3*, 512-517.

[8] Goldstein, M. (1975b), "A Note on Some Bayesian Nonparametric Estimates", *Ann. Stat. 3*, 736-740.

[9] Mouchart, M. and L. Simar (1980), "Least-Squares Approximations in Bayesian Analysis" (with Discussions), *Bayesian Statistics*, edited by J.M. Bernardo, M.H. De Groot, D.V. Lindley and A.F.M. Smith, Valencia University Press.

[10] Mouchart, M. and L. Simar (1982), "Theory and Application of Least Squares Approximations in Bayesian Analysis", CORE Discussion Paper n° 8207, U.C.L., Louvain-la-Neuve, Belgium.

[11] Naylor, J. (1982), "Approximate inferences for a mixture distribution", in preparation.

[12] Relles, D.A. and W.H. Rogers (1977), "Statisticians are Fairly Robust Estimators of Location", *J. Am. Stat. Assoc. 72*, 107-111.

[13] Rolin, J.M. (1982), "Non Parametric Bayesian Statistics : A Stochastic Process Approach, CORE Discussion Paper n° 8225, U.C.L., Louvain-la-Neuve, Belgium.

[14] Simar, L. (1982), "A Survey of Bayesian Approaches to Nonparametric Statistics", to appear in *Math. Operationsforsch. Stat., Ser. Stat.*

[15] Smith, A.F.M. (1977), "Bayesian Statistics and Efficient Information Processing Constrained by Probability Models", in *Decision Making and Change in Human Affairs*, edited by H. Jungerman and G. de Zeeuw, D. Reidel Publishing Company, Dordrecht.

[16] Smith, A.F.M. and U.E. Makov (1978), "A Quasi-Bayes Sequential Procedure for Mixtures", *J.R. Stat. Soc., Ser. B 40*, 106-112.

[17] Stone, M. (1963), "Robustness of Non-Ideal Decision Procedures", *J. Am. Stat. Assoc. 58*, 480-486.

BAYESIAN APPROACHES TO OUTLIERS AND ROBUSTNESS

by

A.F.M. Smith

Department of Mathematics
University of Nottingham

Abstract

A general, Bayesian approach to robustification via model elaboration is introduced and discussed. The approach is illustrated by considering the elaboration of standard models to incorporate the possibility of non-standard distributional shapes or of individual aberrant observations (outliers). Influence functions are then considered from a Bayesian point of view and an approach to robust time series analysis is outlined.

Key-words : Bayesian robustness, Outliers, Time series

AMS/MOS : Primary 62A15, Secondary 62G35, 62M10

1. INTRODUCTION

We shall assume throughout this paper that our concern is with providing operational, statistical methodology for contexts in which - either as a result of preliminary data analysis, or on the basis of conventional wisdom, or through consideration of mathematical tractability - a more-or-less standard parametric form is currently proposed as the basis for statistical analysis. When it is known that actual departures from the assumptions underlying the standard form could cause inferences based on the latter to be badly misleading, there is a need for model robustification, which, according to Box (1980), consists of "judicious and grudging elaboration of the [currently proposed] model to ensure against particular hazards".

If we denote the currently proposed model by M, and the inferential conclusions which follow from M, for a given set of data, D, by C, then the idea of an elaborated model, EM, leading to a range of possible conclusions, RC, can be represented schematically by Figure 1.

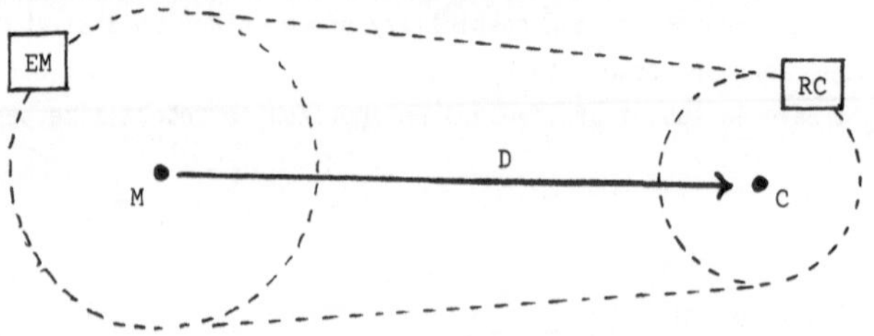

Figure 1: Model elaboration leading to a range of possible conclusions

With reference to Figure 1, if RC is "small" when EM is an "interestingly large" neighbourhood of M, we conclude that the conjunction of M and D constitutes a "robust" inferential context. If RC is "large", however, we have a serious problem of sensitivity of conclusions to assumptions and require as much feedback as possible from D regarding the relative plausibilities of the alternatives contained within EM.

In any case, we require a framework which enables us to study the "mapping" from EM to RC for interesting choices of the former, in relation to a given standard model M. In this paper, we shall largely concentrate on contexts where M corresponds to either the univariate location-scale model, the general linear model, the multivariate location-scale model or the dynamic linear model

(Kalman Filter), all with Gaussian distributional assumptions, and EM corresponds to model elaborations which include the possibility of one or more outliers, or non-Gaussian distributions.

In general, of course, the very notion of a "currently proposed, standard model" is subject to continual revision and cannot be other than a momentarily fixed frame in the moving sequence of model building proposals and checks. Strictly speaking, therefore, EM's should themselves be elaborated to EEM s , and so on. However, real investigations do seem to proceed typically by just a few steps at a time and so, at any given stage, the approach summarized in Figure 1 has considerable pragmatic appeal.

2. BAYESIAN INFERENCE FOR AN ELABORATED MODEL

To convey the general flavour of the Bayesian analysis of an elaborated model with the minimum of notational complication, let us suppose that the currently proposed standard parametric model is represented, for data $\underset{\sim}{x}$, by the density $p(\underset{\sim}{x}|\underset{\sim}{\theta})$, where $\underset{\sim}{\theta}$ is an unknown parameter vector, and that the elaborated neighbourhood of models is represented by the family of densities $\{p(\underset{\sim}{x}|\underset{\sim}{\theta},\lambda),\ \lambda \in \Lambda\}$, where Λ is some form of "labelling set", with $p(\underset{\sim}{x}|\underset{\sim}{\theta}) = p(\underset{\sim}{x}|\underset{\sim}{\theta},\lambda_0)$ for some $\lambda_0 \in \Lambda$.

For example, if $\underset{\sim}{\theta}$ consists of a single unknown location parameter, a very simple elaboration of the standard assumption of a Gaussian distribution (with unit scale) would be to take $\Lambda = \{\lambda_0 = \text{Gaussian},\ \lambda_1 = \text{Uniform},\ \lambda_2 = \text{Double Exponential}\}$. As another example, we might consider elaboration to the family t-densities, labelled by λ^{-1}= degrees of freedom. With $\Lambda = [0,1]$, we have a closed interval of elaborated models, running from the Gaussian distribution ($\lambda = 0$) to the Cauchy distribution ($\lambda = 1$).

Applying the usual Bayesian paradigm to the elaborated framework, we see that inferences about $\underset{\sim}{\theta}$ are summarized by

$$p(\underset{\sim}{\theta}|\underset{\sim}{x}) = \int p(\underset{\sim}{\theta}|\underset{\sim}{x},\lambda)p(\lambda|\underset{\sim}{x})d\lambda, \tag{2.1}$$

where

$$p(\underset{\sim}{\theta}|\underset{\sim}{x},\lambda) \propto p(\underset{\sim}{x}|\underset{\sim}{\theta},\lambda)p(\underset{\sim}{\theta}|\lambda) \tag{2.2}$$

$$p(\lambda|\underset{\sim}{x}) \propto p(\underset{\sim}{x}|\lambda)p(\lambda) \tag{2.3}$$

and

$$p(\underset{\sim}{x}|\lambda) \propto \int p(\underset{\sim}{x}|\underset{\sim}{\theta},\lambda)p(\underset{\sim}{\theta}|\lambda)\, d\theta. \tag{2.4}$$

As was pointed out by Box and Tiao (1964), the individual elements appearing in (2.1) - (2.4) bring out all the relevant features of the approach of Figure 1:

(i) for given $p(\underset{\sim}{\theta},\lambda) = p(\underset{\sim}{\theta}|\lambda)p(\lambda)$ and $\underset{\sim}{x}$, the range of possible inferences about $\underset{\sim}{\theta}$ (RC) corresponding to different choices of $\lambda \in \Lambda$ (EM) is described by $p(\underset{\sim}{\theta}|\underset{\sim}{x},\lambda)$ considered as a function of $\lambda \in \Lambda$;

(ii) since λ labels the form of departure from the assumptions underlying the model M ($\lambda = \lambda_0$), the form of $p(\lambda)$ can be chosen to reflect whatever beliefs (e.g. pessimistic or neutral) about actual departures from M are held, or are of interest to study;

(iii) for any given $p(\lambda)$, and specification of $p(\underset{\sim}{\theta}|\lambda)$, the density $p(\lambda|\underset{\sim}{x})$ provides information about the relative plausibilities of the alternative elaborations contained in EM.

Imaginative displays of combinations of these elements for various choices of $p(\underset{\sim}{\theta}|\lambda)$ and $p(\lambda)$ provide a very comprehensive basis for inference about $\underset{\sim}{\theta}$, or its components. If simple summary estimates and measures of uncertainty are required, the posterior mean, $E(\underset{\sim}{\theta}|\underset{\sim}{x})$, and covariance matrix, $V(\underset{\sim}{\theta}|\underset{\sim}{x})$ can be quoted.

Considering, for simplicity, the case of a single parameter θ, we have

$$E(\theta|\underset{\sim}{x}) = \int E(\theta|\underset{\sim}{x},\lambda)p(\lambda|\underset{\sim}{x})d\lambda = E[E(\theta|\underset{\sim}{x},\lambda)|\underset{\sim}{x}] \tag{2.5}$$

$$V(\theta|\underset{\sim}{x}) = E[V(\theta|\underset{\sim}{x},\lambda)|\underset{\sim}{x}] + V[E(\theta|\underset{\sim}{x},\lambda)|\underset{\sim}{x}]. \tag{2.6}$$

The posterior mean, (2.5), is simply an <u>adaptive weighted average</u> (with respect to $p(\lambda|\underset{\sim}{x})$) of the posterior mean estimates of θ corresponding to specific choices of λ. Estimates of θ corresponding to plausible (in the light of the data) elaborations in EM receive due weight, whereas estimates corresponding to very implausible elaborations are discounted.

The form of the measure of uncertainty about θ, (2.6), is an interesting combination of two terms: the first measures the average <u>"within-model" uncertainty</u> (for alternative model elaborations in EM), whereas the second term measures the <u>"between-model" uncertainty</u> in EM (by assessing the variation between point estimates of θ for different values of λ, taking into account their plausibilities as measured by $p(\lambda|\underset{\sim}{x})$). In cases where several model elaborations are plausible, given $\underset{\sim}{x}$, but lead to very different estimates of θ, the second term may provide the major contribution to overall uncertainty.

3. MODEL ELABORATION TO INCLUDE NON-GAUSSIAN DISTRIBUTIONS

3.1 Some possible elaborations for the location-scale case

In this section we shall suppose that $\underset{\sim}{\theta} = (\mu,\sigma)$, representing location and scale, respectively, and that, given $\underset{\sim}{\theta}$, under the standard model M observations are assumed independent with Gaussian distributions. We begin by listing some possible model elaborations for this case, together with their motivations.

1) A Gaussian/Double Exponential/Uniform family.

In situations where sample size is rather small (less than 15, say), elaboration from the Gaussian model to a large, rich family of alternatives seems rather pointless, since, on the basis of such a small data set, we cannot hope for very fine discrimination in EM. On the other hand, some kind of elaboration that would at least enable us to learn about departures in tail-behaviour from normality (in either a light - or heavy-tailed direction) does seem worthwhile. If we continue to assume symmetry, the choice $\Lambda = \{\lambda_G, \lambda_D, \lambda_U\}$, using an obvious notation, provides one such elaboration. Further discussion of this choice is given in Section 3.2.

2) A Gaussian/Double Exponential/Uniform/Right-Exponential/Left-exponential family.

Again, considering situations where sample size is small, but wishing to allow for the possibility of skew departures from normality (in addition to possible symmetric departures in tail behaviour), a simple family containing "token" representative densities corresponding to light- and heavy-tail behaviour, skewness to the right and left is given by $\Lambda = \{\lambda_G, \lambda_D, \lambda_U, \lambda_R, \lambda_L\}$, where

$$p(x|\mu,\sigma,\lambda_R) = \frac{1}{e\sigma}\exp\left\{-\left(\frac{x-\mu}{\sigma}\right)\right\}, \qquad x \geq \mu - \sigma$$

and λ_L corresponds to the "reflection" of λ_R about $\mu - \sigma$.

3) The exponential power family.

If we take

$$p(x|\mu,\sigma,\lambda) \propto \sigma^{-1}\exp\left\{-c(\lambda)\left|\frac{x-\mu}{\sigma}\right|^{\left(\frac{2}{1+\lambda}\right)}\right\}, \qquad x \in \mathbb{R},$$

with $\Lambda = (-1,+1]$, we obtain a family of densities ranging from the uniform $(\lambda \to -1)$, through the normal $(\lambda = 0)$ to the double exponential $(\lambda = +1)$, where λ may be viewed as a measure of kurtosis.

Detailed illustration of the application of the approach of Section 2 using this family is given in Box and Tiao (1973).

4) The t-family.

If we consider the usual t-family of densities with location μ and scale σ, together with λ^{-1} = degrees of freedom, then the elaborated model consisting of $\Lambda = [0,1]$ contains a range of models from normality ($\lambda = 0$) through increasingly heavier-tailed t-alternatives to the Cauchy distribution ($\lambda = 1$).

This family was used in a study by Relles and Rogers (1977) to produce robust location estimates, based on (2.5), which were shown to have good properties (in a comparative sampling study).

5) The contaminated-normal family

With $\lambda = (\varepsilon, k)$, $0 < \varepsilon < 1$, $k > 1$, the family of densities

$$p(x|\mu,\sigma,\lambda) = (1-\varepsilon)\phi(x|\mu,\sigma) + \varepsilon\phi(x|\mu,k\sigma), \quad x \in \mathbb{R},$$

where ϕ denotes the normal density, represents an ε-contamination of M by an inflated-variance normal component. If we take, for the purpose of illustration,

$$\Lambda = \{(0,0.1] \times [2,3]\},$$

we have an EM which guards against up to 10% contamination by observations whose standard derivations are between 2 and 3 times the norm.

This family is somewhat inconvenient for routine use because of the two-dimensional nature of λ. However, the following family provides a satisfactory alternative.

6) The Huber family

A family of densities well-known from the classical theory of M-estimators (Huber, 1964) is defined by

$$p(x|\mu,\sigma,\lambda) \propto \exp\left\{-u\left(\frac{x-\mu}{\sigma}\right)\right\}, \quad x \in \mathbb{R},$$

where

$$u(t) = \begin{cases} \frac{1}{2}t^2, & |t| < \lambda \\ \lambda|t| - \frac{1}{2}\lambda^2, & |t| \geq \lambda. \end{cases}$$

As $\lambda \to \infty$, the density tends to the normal form; as $\lambda \to 0$, it resembles a double exponential. For intermediate values of λ, we have a density with normal centre and exponential tails.

To motivate this family as an EM, we could appeal directly to its relationship with M-estimation. From this perspective, and invoking the usual invariant location-scale prior form, the form (2.5) applied marginally to μ

turns out to be (approximately) an adaptive weighted average of M-estimators corresponding to particular choices of λ.

As a more "practical" motivation, we recall from Tukey (1960) that samples from contaminated normal distributions give rise to frequency curves which are remarkably similar in shape to forms with normal centres and exponential tails. In order to understand better the form of many-one mapping $(\varepsilon,k) \to \lambda$ which would enable us to make a choice of Λ corresponding to an interesting range of (ε,k) in 5) above, we could investigate, using, for example, the Kullback-Leibler information measure, which member of the family was "closest" to a specified member of the contaminated normal family. Such a study has been carried out by Naylor and Smith (1981) and establishes that the use of 6) with $\Lambda = [1.2, 2.5]$ provides an excellent pragmatic EM substitute for the rather intractable EM based on 5) with $.02 < \varepsilon < .1$ and $2 < k < 3$.

3.2 Bayesian robust estimation using $\Lambda = \{\lambda_G, \lambda_D, \lambda_U\}$.

Spiegelhalter (1981) has considered the use of the marginal posterior mean for μ (corresponding to (2.5), based on the EM $\Lambda = \{\lambda_G, \lambda_D, \lambda_U\}$, with $p(\lambda) = \frac{1}{3}$, $p(\mu,\sigma|\lambda) \propto \sigma^{-1}$, $\lambda \in \Lambda$) as an adaptive robust estimator of location.

The marginal likelihoods for the three shapes, $p(\underset{\sim}{x}|\lambda)$, can all be calculated explicitly for this choice of Λ (see Uthoff, 1970, 1973) and, for a sample $\underset{\sim}{x} = (x_1, \ldots, x_n)$, are given by:

$$p(\underset{\sim}{x}|\lambda_G) = \Gamma\left(\frac{n-1}{2}\right) 2^{-1} n^{-\frac{1}{2}} \left[\pi s^2(n-1)\right]^{-\frac{1}{2}(n-1)}, \tag{3.1}$$

where $s^2 = (n-1)^{-1} \sum (x_i - \bar{x})^2$,

$$p(\underset{\sim}{x}|\lambda_D) = \Gamma(n-1) 2^{-(n+1)} V_m^{1-n} \sum u_j^{-1}, \tag{3.2}$$

where

$$V_j = \sum_i (x_{[i]} - x_{[j]}), \qquad m = [(n+1)/2]$$

$$u_j = \begin{cases} 4(V_j/V_m)^{n-1}(j - \frac{n}{2})(\frac{n}{2} + 1 - j) & j \neq \frac{n}{2}, \frac{n}{2}+1 \\ 4\left[1 + (n-1)V_m^{-1}(x_{[m+1]} - x_{[m]})\right]^{-1} & j = \frac{n}{2}, \frac{n}{2}+1 \end{cases}$$

and

$$p(\underset{\sim}{x}|\lambda_U) = n^{-1}(n-1)^{-1}(x_{[n]} - x_{[1]})^{-(n-1)} \tag{3.3}$$

The forms (3.1), (3.2), (3.3) provide the basis for the calculation of posterior weights on the three alternatives in EM (see (2.3)), so that, for example, with $p(\lambda) = \frac{1}{3}$,

$$p(\lambda_G|\underset{\sim}{x}) = \left[1 + \frac{p(\underset{\sim}{x}|\lambda_D)}{p(\underset{\sim}{x}|\lambda_G)} + \frac{p(\underset{\sim}{x}|\lambda_U)}{p(\underset{\sim}{x}|\lambda_G)}\right]^{-1} .$$

These posterior weights determine the adaptive nature of the posterior mean estimator based on Λ, since

$$E(\mu|\underset{\sim}{x}) = \hat{\mu}_G\, p(\lambda_G|\underset{\sim}{x}) + \hat{\mu}_D\, p(\lambda_D|\underset{\sim}{x}) + \hat{\mu}_U\, p(\lambda_U|\underset{\sim}{x}), \tag{3.4}$$

where, for example, $\hat{\mu}_G = E(\mu|\underset{\sim}{x},\lambda_G)$. It is therefore interesting to note that the ratio $p(\underset{\sim}{x}|\lambda_i)/p(\underset{\sim}{x}|\lambda_G)$ is the uniformly most powerful location and scale invariant test statistic between the shapes λ_i and λ_G. The forms of the adaptive weights thus use the data in a highly efficient way.

In fact, $\hat{\mu}_G = \bar{x}$, $\hat{\mu}_U = \frac{1}{2}(x_{[n]}+x_{[1]})$ and $\hat{\mu}_D \approx x_{[m]}$, so that (3.4) has (approximately) the form of an adaptive weighted average of the sample mean, mid-range and median. In a detailed sampling study, Spiegelhalter shows that, using conventional criteria, (3.4) provides an efficient form of robust estimator for small samples against a range of true underlying sample distributions and performs well in comparison with several classically proposed forms of M-estimator and adaptive trimmed mean.

As an illustration of the general approach given in Section 2, Spiegelhalter (1978) analysed a data set taken from Stigler (1977) and consisting of 23 measurements (given in Table 1) made by Cavendish in 1798 in an experiment to determine the density of the earth. The configuration of the data is displayed in Figure 2.

Table 1. Cavendish Data

5.10	5.27	5.29	5.29	5.30	5.34
5.34	5.36	5.39	5.42	5.44	5.46
5.47	5.53	5.57	5.58	5.62	5.63
5.65	5.68	5.75	5.79	5.85	

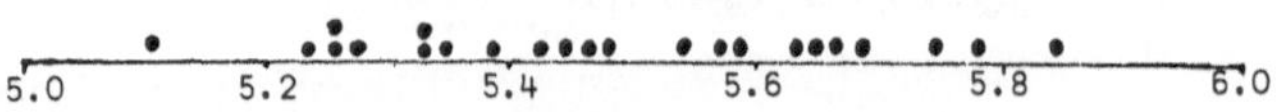

Figure 2. Dot diagram of the Cavendish data.

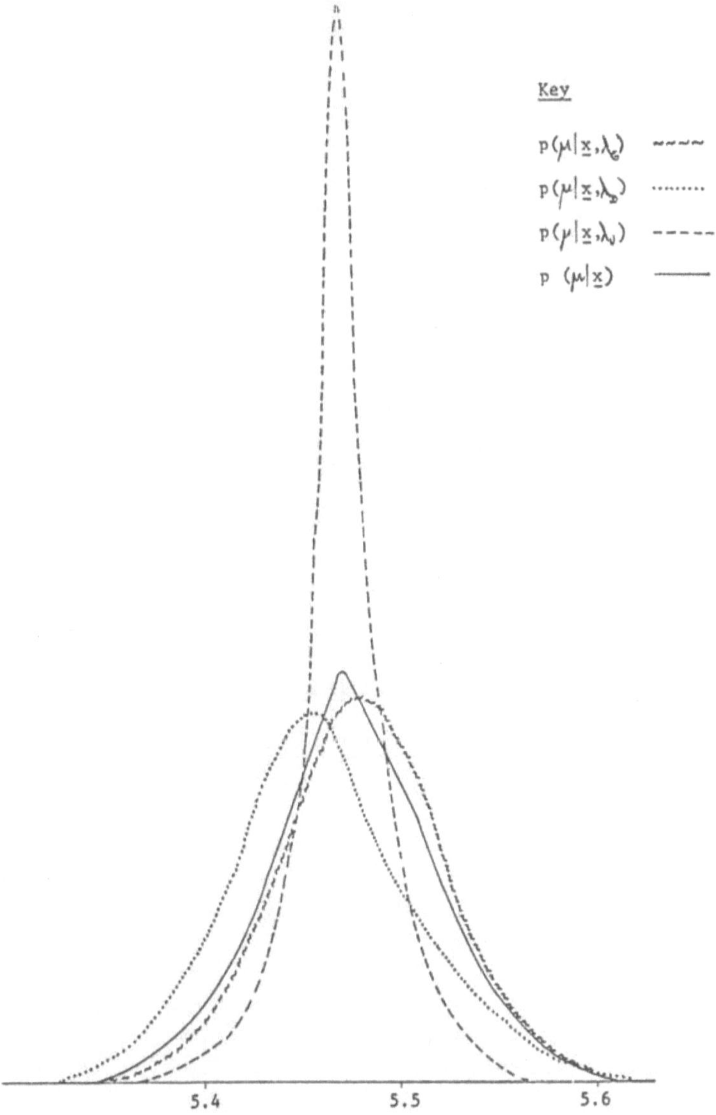

Figure 3: Location Parameter Inference for Cavendish Data

The results obtained by applying the approach of Section 2 (using the prior specifications described above) are summarized in Table 2.

Table 2. Results for the Cavendish Data using $\Lambda = \{\lambda_G, \lambda_D, \lambda_U\}$.

	λ_G	λ_D	λ_U
$P(\lambda \mid \underset{\sim}{x})$	.810	.132	.058
$E(\mu \mid \underset{\sim}{x}, \lambda)$	5.484	5.464	5.475
$V(\mu \mid \underset{\sim}{x}, \lambda)$	.0017	.0022	.0007

The overall inferences, summarized by (2.5), (2.6), are

$$E(\mu \mid \underset{\sim}{x}) = 5.480, \quad V(\mu \mid \underset{\sim}{x}) = .0018 .$$

In this case, quite strong support is given to the Gaussian distribution. Figure 3 displays the posterior densities, $p(\mu \mid \underset{\sim}{x}, \lambda)$, for each element of Λ, together with the marginal posterior $p(\mu \mid \underset{\sim}{x})$.

3.3 Discussion

It is natural to enquire about the status of the analysis of Section 2 when the actual underlying distribution does not belong to the chosen EM, characterized by Λ. A detailed discussion is given in Spiegelhalter (1981), but the basic behaviour of the posterior inferences may be summarized as follows.

If the EM is represented by $p(\underset{\sim}{x} \mid \underset{\sim}{\theta}, \lambda)$, $\lambda \in \Lambda$, and the true density is in fact $p_0(\underset{\sim}{x})$, it can be shown (using a theorem of Berk, 1966) that (subject to some weak regularity conditions) the posterior density $p(\underset{\sim}{\theta}, \lambda \mid \underset{\sim}{x})$ will concentrate (with probability one, as $n \to \infty$) on the subset of $\Theta \times \Lambda$ for which $\int \log p(x \mid \underset{\sim}{\theta}, \lambda) p_0(x)\, dx$ is maximized. Alternatively, we can rephrase this by saying that the posterior concentrates on the subset of elements of $\Theta \times \Lambda$ which minimize the Kullback-Leibler directed divergence between $p_0(x)$ and $p(x \mid \underset{\sim}{\theta}, \lambda)$. Typically, this subset consists of a point $(\underset{\sim}{\theta}^*, \lambda^*)$ in $\Theta \times \Lambda$ and we shall assume this in what follows.

The asymptotic posterior distribution of μ resulting from an EM characterized by Λ is then that corresponding to the posterior conditioned on

λ^*. To see this, we note from Berk (1966) that $\lim\sup p(\lambda|\underset{\sim}{x})^{\frac{1}{n}} < 1$ when $\lambda \neq \lambda^*$ and hence $p(\lambda|\underset{\sim}{x})$ is $O_p(C_\lambda^n)$ for some $C_\lambda < 1$. On the other hand, $\sup p(\mu|\underset{\sim}{x},\lambda)$ (wrt μ) is in general $O(n^k)$, where $k = \frac{1}{2}$, typically (although $k = 1$ for λ_U). It follows that

$$\begin{aligned} p(\mu|\underset{\sim}{x}) &= \sum_\lambda p(\mu|\underset{\sim}{x},\lambda)p(\lambda|\underset{\sim}{x}) \\ &\xrightarrow{p} p(\mu|\underset{\sim}{x},\lambda^*) \end{aligned}$$

uniformly in μ.

Using techniques similar to Birnbaum and Miké (1970), it can also be shown that $E(\mu|\underset{\sim}{x})$ has the same asymptotic sampling distribution as $E(\mu|\underset{\sim}{x},\lambda^*)$.

These various theoretical considerations, and others (given in more detail by Spiegelhalter, 1981), enable us to examine to some extent the suitability of a particular proposed EM (involving some considerations of operational tractability) in the light of realistic assessments of plausible actual (but perhaps intractable) underlying distributions.

The particular Λ considered in the previous section was originally proposed as a simple case which enabled some theoretical study of the approach to be made. However, even this simple EM displays remarkable adaptive properties and strongly suggests that richer families of shapes (including some of those outlined in Section 3.1) will often provide a very satisfactory framework for inference.

For further related analyses based on $\Lambda = \{\lambda_G,\lambda_D,\lambda_U\}$ and $\Lambda = \{\lambda_G,\lambda_D,\lambda_U,\lambda_R,\lambda_L\}$, see Spiegelhalter (1977, 1980).

4. MODEL ELABORATION TO INCLUDE OUTLIERS

4.1 Introduction

We consider first, for simplicity, the univariate location-scale situation in which the standard model M assumes observations to be independent and described by the Gaussian density $\phi(x|\mu,\sigma)$. Typical forms of model elaboration to include the possibility of outliers (or aberrant observations) assume that "good" observations are described by M, whereas "bad" observations follow an alternative Gaussian density $\phi_\delta(x|\mu,\sigma)$, this latter being assumed to be one of:

$\phi_\delta(x|\mu,\sigma) = \phi(x|\mu+\delta,\sigma)$, the <u>location-shift</u> model or

$$\phi_\delta(x|\mu,\sigma) = \phi(x|\mu,\delta\sigma), \quad \text{the } \underline{\text{inflated-variance}} \text{ model.}$$

Further scope in the precise form of elaboration used is provided by the choice of range of number of potential outliers to be entertained. In addition, models differ in whether δ is assumed fixed, or varying from one aberrant observation to another, and in this case, having possibly different signs. Various such elaborations are considered in detail by Box and Tiao (1968), Abraham and Box (1978), Guttman, Dutter and Freeman (1978) and Freeman (1981).

Given data $\underset{\sim}{x} = (x_1,\ldots,x_n)$, one way of describing the elaboration to take account of possible outliers is by a "label" set Λ, consisting of all subsets of the observation labels $\{1,\ldots,n\}$ which might be regarded as "bad" observations. Thus if at most one outlier is considered feasible, we might take $\Lambda = \{\emptyset,\{1\},\ldots,\{n\}\}$. If up to two outliers are considered feasible, Λ might consist of the empty set $\emptyset$, all single subscript subsets and all subsets consisting of two subscripts, and so on.

If μ is the parameter of interest, we then have, from (2.1),

$$p(\mu|\underset{\sim}{x}) = \sum_{\lambda\in\Lambda} p(\mu|\underset{\sim}{x},\lambda)p(\lambda|\underset{\sim}{x}) \tag{4.1}$$

where $p(\mu|\underset{\sim}{x},\lambda)$ now describes the posterior density for μ, having assumed those elements of $\underset{\sim}{x}$ with labels in λ to be outliers, and $p(\lambda|\underset{\sim}{x})$ provides the posterior probability that this particular subset of observations does, in fact, consist of outliers.

One problem with such an approach is the computational explosion implicit in (4.1) (for example, allowing for up to 5% outliers in a sample of size 40 gives a Λ with 821 basic elements). Other problems are indicated by Freeman (1981).

In Section 4.2, we shall outline a modified approach to outlier elaborations which avoids many of the difficulties just referred to. In subsequent sections we summarize its detailed development for some specific models.

4.2 A modified approach to outlier elaborations

We shall present the basic ideas in the context of a univariate location-scale problem, where we assume that there is at most one bad observation described by $\phi(x|\mu+\delta,\sigma)$, where $\delta > 0$ (which we shall refer to as an upper outlier; $\delta < 0$ defines a lower outlier).

The events "there is one bad observation" and "the ith observation is bad, all the others are good" are denoted by M_1 and $M_1(i)$, respectively.

Defining τ to be a permutation on $\{1,\dots n\}$ such that

$$\tau(i) = j \iff x_i = x_{(j)},$$

where $x_{(1)} < \dots < x_{(n)}$ are the ordered observations, we further denote the event "$\tau^{-1}(n) = i$" by C_i.

Suppose now that $\underset{\sim}{x}$ is such that $\tau^{-1}(n) = i^*$. We shall assume that the specification of $p(\delta)$, our prior knowledge about the magnitude of the slippage, is such that

$$P(M_1(i) \cap C_i') \approx 0. \tag{4.2}$$

This is equivalent to saying that the slippage mechanism producing the aberrant observation is such that with high probability the bad observation will turn out to be the largest.

It then follows that

$$P(M_1|\underset{\sim}{x}) = \sum_i P(M_1(i)|C_i,\underset{\sim}{x})P(C_i|\underset{\sim}{x}) + \sum_i P(M_1(i) \cap C_i'|\underset{\sim}{x}) \approx P(M_1(i^*)|\underset{\sim}{x}), \tag{4.3}$$

using (4.2), and by noting that (since $\underset{\sim}{x} \Rightarrow \tau^{-1}(n) = i^*$)

$$P(C_i|\underset{\sim}{x}) = \begin{cases} 1 & \text{if } i = i^* \\ 0 & \phantom{\text{if }} i \neq i^*. \end{cases}$$

In a similar way, if we consider the possibility of two upper outliers and $\underset{\sim}{x}$ is such that $\tau^{-1}(n) = i^*$, $\tau^{-1}(n-1) = i^{**}$, an obvious extension of the above argument and notation establishes that

$$P(M_2|\underset{\sim}{x}) \approx P(M_2(i^*,i^{**})|\underset{\sim}{x}). \tag{4.4}$$

The argument carries over to consideration of upper and lower outliers and means, in effect, that we are adding to the "bad" observation assumption, $\phi_\delta(x|\mu,\sigma)$, a form of specification of $p(\delta)$ which implies that only "outlying" observations deserve consideration as potentially aberrant members of the sample. The number of terms in (4.1) is then dramatically reduced (cf. (4.3) and (4.4)). The assumption (4.2) (and its obvious extensions) means that when it comes to specifying a model with r_1 lower outliers and r_2 upper outliers (denoted by $M(r_1,r_2)$) we need only consider the case where observations with labels in $\{\tau^{-1}(j),\ 1 \le j \le r_1\}$ correspond to $\phi(x|\mu+\delta,\sigma)$, with $\delta < 0$, and observations with labels in $\{\tau^{-1}(j),\ n-r_2+1 \le j \le r\}$ correspond to $\phi(x|\mu+\delta,\sigma)$ with $\delta > 0$: the remaining observations correspond to $\phi(x|\mu,\sigma)$.

If we denote by $\underset{\sim}{\Delta}$ the parametrization of the δ-shifts (which may be allowed to vary from one aberrant observation to another), the full model elaboration consists in specifying

$$p(\underset{\sim}{x}|\mu,\sigma,\underset{\sim}{\Delta},M(r_1,r_2)), \tag{4.5}$$

$$p(\mu,\sigma,\underset{\sim}{\Delta}|M(r_1,r_2)), \tag{4.6}$$

$$p(M(r_1,r_2)). \tag{4.7}$$

Combining (4.5) and (4.6) gives $p(\underset{\sim}{x}|M(r_1,r_2))$, which, together with (4.7), provides $p(M(r_1,r_2)|\underset{\sim}{x})$. From (4.5) and (4.6) we also obtain $p(\mu|\underset{\sim}{x},M(r_1,r_2))$, and combining these two latter factors provides overall inferences about μ. In particular, an estimate of μ is given by

$$E(\mu|\underset{\sim}{x}) = \sum_{(r_1,r_2)} E(\mu|\underset{\sim}{x},M(r_1,r_2))p(M(r_1,r_2|\underset{\sim}{x}), \tag{4.8}$$

a much simplified version of the adaptive weighted average (4.1).

4.3 Univariate location-scale elaborations

Denoting the ordered version* of $\underset{\sim}{x}$ by $\underset{\sim}{x}^*$, inferences derived on the basis of the approach given in Section 4.2 will proceed as if for model $M(r_1,r_2)$ (with possibly different δ-shifts for individual aberrant observations) we assume

$$\begin{aligned} \underset{\sim}{x}^* &\sim N(\underset{\sim}{A}_1\underset{\sim}{\theta}_1,\underset{\sim}{C}_1) \\ \underset{\sim}{\theta}_1 &\sim N(\underset{\sim}{A}_2\underset{\sim}{\theta}_2,\underset{\sim}{C}_2) \end{aligned} \tag{4.9}$$

with

$$\underset{\sim}{A}_1 = \begin{bmatrix} & \vdots & \underset{\sim}{I}_{r_1} & \underset{\sim}{0} \\ \underset{\sim}{1}_n & \vdots & \underset{\sim}{0} & \underset{\sim}{0} \\ & \vdots & 0 & \underset{\sim}{I}_{r_2} \end{bmatrix} \quad \underset{\sim}{C}_1 = \sigma^2\underset{\sim}{I}_n, \qquad \underset{\sim}{\theta}_1 = \begin{bmatrix} \mu \\ \delta_1 \\ \vdots \\ \delta_{r_1} \\ \delta_{n-r_2+1} \\ \vdots \\ \delta_n \end{bmatrix},$$

$$\underset{\sim}{A}_2 = \begin{vmatrix} 1 & 0 \\ \underset{\sim}{0} & -\underset{\sim}{1}_{r_1} \\ 0 & \underset{\sim}{1}_{r_2} \end{vmatrix}, \qquad \underset{\sim}{C}_2 = \sigma^2\begin{vmatrix} k^2 & 0 \\ 0 & \eta^2\underset{\sim}{I}_{r_1+r_2} \end{vmatrix}, \quad \underset{\sim}{\theta}_2 = \begin{vmatrix} \mu_0 \\ \delta_0 \end{vmatrix},$$

with $\delta_0 > 0$.

Detailed analysis of the elaborated model follows straightforwardly from (4.9), using the general results of Lindley and Smith (1972). In particular, in the case where k^{-2} and η^{-2} are small, we have $E(\mu|\underset{\sim}{x},M(r_1,r_2))$ approximately

---

* By "ordered version", we mean that the observations have been partially reordered so that the r_1 lower outliers are written first, and the r_2 upper outliers are written last.

equal to the trimmed mean obtained by removing the r_1 lower and r_2 upper outliers. The form of $E(\mu|\underset{\sim}{x})$ implied by (4.8) is then an adaptive weighted average of trimmed means, the adaptive weights being the posterior probabilities of the various particular assumptions about the numbers of outliers.

Further insight into these weights is provided by noting that if M_0 denotes the assumption of no outliers, and $\underset{\sim}{A}_1$ above is written in the form $[\underset{\sim}{1}_n : \underset{\sim}{A}]$, then the posterior to prior odds ratio

$$[p(M_0|\underset{\sim}{x})/p(M(r_1,r_2)|\underset{\sim}{x})]/[p(M_0)/p(M(r_1,r_2))]$$

is approximately equal to

$$|n^2\underset{\sim}{I}_r|^{\frac{1}{2}}|\underset{\sim}{A}^T\underset{\sim}{A}|^{\frac{1}{2}}\left|1+\frac{(r_1+r_2)}{n-(r_1+r_2+1)}F\right|^{-\frac{n}{2}},$$

where $r = r_1 + r_2$ and F is the usual F statistic for testing M_0 against $M(r_1,r_2)$; see, for example, Smith and Spiegelhalter (1981). We see, therefore, that the adaptive weights are monotone decreasing functions of conventional statistics for testing for outliers.

The inflated-variance elaboration can also be represented within the framework (4.9), but further details will not be given here. Instead, we shall briefly comment on a multivariate form of the inflated variance model.

4.4 A multivariate elaboration

In the univariate case, the approach adapted to obtaining a computationally tractable form of elaboration exploited the obvious ordering of the data, together with a prior specification which ensured that an observation could only acquire non-negligile posterior probability of being an outlier if, in fact, it lay on the "outside" of the sample. In multivariate contexts, there is, of course, no such natural ordering and so a more general approach is required.

If $\underset{\sim}{\psi}$ denotes the vector of all parameters occuring in the basic model, and $\underset{\sim}{x}(s)$ denotes the elements of $\underset{\sim}{x}$ whose labels occur in the set $s \subseteq \{1,...,n\}$, then

$$p(\underset{\sim}{x}(s)|\underset{\sim}{x}(s')) = \int p(\underset{\sim}{x}(s)|\underset{\sim}{\psi})p(\underset{\sim}{\psi}|\underset{\sim}{x}(s'))\,d\underset{\sim}{\psi} \qquad (4.10)$$

provides the predictive density for $\underset{\sim}{x}(s)$ given $\underset{\sim}{x}(s')$, where s' denotes the complement of s. The quantity defined by (4.10) provides a measure of the "surprisingness" of observations $\underset{\sim}{x}(s)$ in the light of observations $\underset{\sim}{x}(s')$ (and the prior specification) and can be used to order individual observations, pairs, triples, etc., on the basis of their "aberrant" nature compared with the other observations.

Taking $N_p(\underset{\sim}{\mu},\underset{\sim}{\Sigma})$ as the basis model for "good" observations and $N_p(\underset{\sim}{\mu},\delta\underset{\sim}{\Sigma})$, $\delta > 1$, as the model for "bad" observations, we shall assume (by analogy with the approach leading to (4.2)) that our prior specification for δ is such that the posterior probability of any set of r observations $\underset{\sim}{x}(t)$, say, being regarded as "bad" is negligible unless, among all subsets of r observations, (4.10) is minimized by s = t. In this case, if M_r denotes the model which asserts that there are r outliers, and we write $\underset{\sim}{x}^* = (\underset{\sim}{x}(t'),\underset{\sim}{x}(t))$, we can proceed as if M_r is modelled by (4.9), with

$$A_1 = \begin{bmatrix} I_p \\ \vdots \\ I_p \end{bmatrix} \begin{matrix}\uparrow \\ n, \\ \downarrow\end{matrix} \quad C_1 = \begin{bmatrix} \Sigma & & & & & \\ & \ddots & & & 0 & \\ & & \Sigma & & & \\ & & & \delta\Sigma & & \\ & 0 & & & \ddots & \\ & & & & & \delta\Sigma \end{bmatrix} \begin{matrix} \uparrow \\ n-r \\ \downarrow \\ \uparrow \\ r \\ \downarrow \end{matrix}, \quad \theta_1 = \mu,$$

$$A_2 = I_p, \qquad C_2 = a^{-1}\Sigma, \qquad \theta_2 = \mu_0,$$

for some a.

Detailed development of this model is straightforward, but algebraically cumbersome. In qualitative terms, posterior inferences for $\underset{\sim}{\Sigma}$, say, consists of adaptive weighted averages of inferences based on particular choices of M_r. It can be proved that, if r = 1, outlier "candidates" (according to the ordering defined by (4.10)) must lie on the convex hull of the sample. If r = 2, candidate pairs of outliers either both lie on the convex hull or one of them lies on the convex hull obtained by deleting the other from the sample; and so on, similarly. This means that estimates of $\underset{\sim}{\Sigma}$ based on our approach have the form of adaptive weighted averages of estimates obtained by successive "peeling" of the layers of the sample "onion". From a computational point of view, there are efficient algorithms for exploring the vertices of the sample convex hull.

A full development of the ideas presented in Section 4 is given in Pettit and Smith (1981).

5. BAYESIAN INFLUENCE FUNCTIONS AND SEQUENTIAL LEARNING

5.1 Influence functions

If we consider, for simplicity, a single observation x from a density $p(x|\theta)$, where θ is a single unknown parameter with prior density $p(\theta)$, then we have, from Bayes theorem,

$$g_\theta(\theta|x) = g_\theta(\theta) + g_\theta(x|\theta),$$

where

$$g_\theta(\theta|x) = -\frac{\partial}{\partial\theta}\log p(\theta|x), \quad g_\theta(\theta) = -\frac{\partial}{\partial\theta}\log p(\theta), \quad g_\theta(x|\theta) = -\frac{\partial}{\partial\theta}\log p(x|\theta).$$

We recognize $g_\theta(x|\theta)$ as the efficient score function (of $p(x|\theta)$ with respect to θ); see Cox and Hinkley (1974). If we term this the likelihood score, and define the other two functions to be the prior and posterior score functions, we can rewrite Bayes theorem in the <u>additive</u> form:

<u>posterior score = prior score + likelihood score</u>.

This form provides a useful and easily understandable basis for the qualitative study of posterior densities for varying x, and for varying combinations of prior and likelihood, and can be helpful in deciding upon suitable choices of EM (recall Figure 1). Ramsay and Novick (1980) have recently made use of this idea.

If we consider, instead, the sensitivity of $p(\theta|x)$ to x, we may define similar score - or <u>influence</u> functions - but now with respect to x. Subject, as above, to obvious regularity conditions, and in an obvious notation,

$$\begin{aligned} g_x(\theta|x) &= g_x(x|\theta) - g_x(x) \\ &= g_x(x|\theta) - E[g_x(x|\theta)|x]. \end{aligned}$$

The function $g_x(\theta|x)$ measures the "influence" of x, and if we wish to develop an EM to protect inferences against the misleading effects of outliers, it will be natural to seek for forms of $p(x|\theta)$, $p(\theta)$ such that the influence of individual observations is bounded, in some sense. This, of course, leads to close connections with the classical theory of robust estimation (the likelihood score can be seen as the influence function for a location M-estimator, in the sense of Hampel, 1968, 1974).

For a sample $\underset{\sim}{x} = (x_1,\ldots,x_n)$, the above ideas extend in an obvious way to give

$$g_{x_j}(\theta|\underset{\sim}{x}) = g_{x_j}(x_j|\theta) - g_{x_j}(\underset{\sim}{x}),$$

where

$$g_{x_j}(\theta|\underset{\sim}{x}) = -\frac{\partial}{\partial x_j}\log p(\theta|\underset{\sim}{x})$$

and

$$g_{x_j}(\underset{\sim}{x}) = -\frac{\partial}{\partial x_j}\log p(\underset{\sim}{x}) = E[g_{x_j}(x_j|\theta)|\underset{\sim}{x}].$$

This makes it clear that the measurement of the influence of an individual observation is relative to other available observations (cf. the discussion of (4.10)).

5.2 Inference for a location parameter

Let us suppose that a single observation x is made of an unknown location parameter θ, so that $x = \theta + \varepsilon$, where ε is a measurement error, giving rise to a symmetric density, $p(x|\theta) = p(x-\theta)$ obeying certain regularity conditions (which will be implicit in what follows). The scale is assumed known and equal to one, for simplicity.

Let us further suppose that the prior density for θ may be assumed to be Gaussian with mean m and variance c^2. Then we have the following:

Theorem (Masreliez, 1975).

If $g_x(x) = -\frac{\partial}{\partial x}\log p(x)$ and $G_x(x) = \frac{\partial}{\partial x}g_x(x)$, where $p(x) = \int p(x-\theta)p(\theta)\,d\theta$ then:

$$\begin{aligned} &(i) \quad E(\theta|x) = m + c^2 g_x(x), \\ &(ii) \quad V(\theta|x) = c^2 - c^4 G_x(x). \end{aligned} \tag{5.1}$$

The proof of (i) follows straightforward by noting that

$$(\theta - m)p(\theta) = -c^2\frac{\partial p(\theta)}{\partial \theta}$$

and then integrating by parts and exchanging the order of integration. The proof of (ii) follows similarly.

The form of (i) is extremely interesting. It shows explicitly how a prior estimate of θ, given by m, is shifted by taking account of the new information x through its marginal influence function. (There are many points of contact here with the work of O'Hagan, 1979.)

Since, broadly speaking, tail behaviour of $p(x-\theta)$ is carried through to p(x) (under convolution with respect to a Gaussian density), the qualitative behaviour of (i) will be essentially determined by the choice of distribution of ε, defining $p(x-\theta)$.

Alternatively, we could work directly in terms of $p(x-\theta)$ by invoking, as an approximation to (i),

$$\begin{aligned} E(\theta|x) &\approx m + c^2\left[-\frac{\partial}{\partial x}\log p(x-m)\right]. \\ &\approx m + c^2 g_x(x|m). \end{aligned} \tag{5.2}$$

Thus, for example, considering some of the families discussed in Section 3.1, we obtain qualitative insight into the behaviour of (5.2) from Figure 4. (See West, 1981, for further details.)

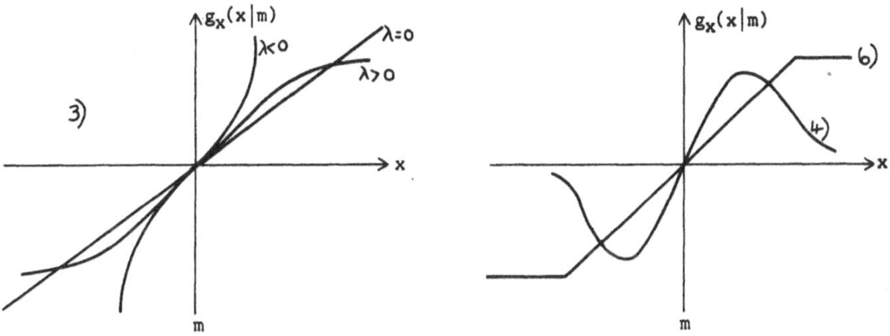

Figure 4: Forms of $g_x(x|m)$ for the exponential power, Huber and t-families.

5.3 Sequential aspects; robust time series analysis

Suppose we now consider a sequential version of the problem considered in Section 5.2, so that $x_n = \theta + \varepsilon_n$, with independent ε_n giving the same form of density $p(x_n|\theta)$ for each n. Then if we assume that the prior for θ - which is now, of course, the posterior based on $(x_1,\ldots,x_{n-1})$ - can be reasonably approximated by an $N(m_{n-1},c_{n-1}^2)$ distribution, (i) in (5.1) can be approximated by

$$E(\theta|x_1,\ldots,x_n) = m_n \approx m_{n-1} + c_{n-1}^2 g_x(x_n|m_{n-1}). \tag{5.3}$$

A similar approximate recursive equation can also be obtained for $c_n^2 = V(\theta|x_1,\ldots,x_n)$, and a robustified fully recursive approximate sequential updating system developed. For further details of the performance of such a system and its extension to the case of unknown scale, see West (1981).

It should now be clear that <u>any</u> form of linear model with arbitrary symmetric additive error, and with an assumed (approximate) Gaussian prior distribution over its location structure, will lead to recursive systems of the type outlined in (5.3). In particular, this applies to the dynamic linear model (introduced by Harrison and Stevens, 1976), which provides scope for modelling a wide range of time series problems. We thus have available a very general Bayesian approach to robust time series analysis. (See, also, Masreliez and Martin, 1977.)

ACKNOWLEDGEMENTS

Detailed investigations of the various model elaborations outlined in the paper have been largely carried out by my past and present research students, first at University College, London and more recently at Nottingham University. It is a pleasure to acknowledge the education I have received from David Spiegelhalter, Michael West, John Naylor and Lawrence Pettit.

REFERENCES

ABRAHAM, B. and BOX, G.E.P. (1978) Linear models and spurious observations. Appl.Statist. 27, 131-8.

BERK, R.H. (1966). Limiting behaviour of posterior distributions when the model is incorrect. Ann.Math.Statist. 37, 51-8.

BIRNBAUM, A. and MIKÉ, V. (1970). Asympototically robust estimators of location. J.Amer.Statist.Ass. 65, 1265-82.

BOX, G.E.P. (1980). Sampling and Bayes' inference in scientific modelling and robustness (with Discussion). J.R. Statist.Soc.A. 143, 383-430.

BOX, G.E.P. and TIAO, G.C. (1964). A Bayesian approach to the importance of assumptions applied to the comparison of variances. Biometrika, 51, 153-67.

BOX, G.E.P. and TIAO, G.C. (1968). A Bayesian approach to some outlier problems. Biometrika, 55, 119-29.

BOX, G.E.P. and TIAO, G.C. (1973). Bayesian Inference in Statistical Analysis. Reading, Mass: Addison-Wesley.

COX, D.R. and HINKLEY, D.V. (1974). Theoretical Statistics. London: Chapman Hall.

FREEMAN, P.R. (1981). On the number of outliers in data from a linear model. In Bayesian Statistics (Ed. Bernardo *et al*). Valencia: University Press.

GUTTMAN, I., DUTTER R. and FREEMAN, P.R. (1978). Care and handling of outliers in the general linear model to detect spurosity — a Bayesian approach. Technometrics 20, 187-93.

HAMPEL, F. (1968). Contributions to the theory of robust estimation. Ph.D. dissertation. University of California, Berkeley.

HAMPEL, F. (1974). The influence curve and its role in robust estimation. J.Amer.Statist.Ass. 69, 383-93.

HARRISON, P.J. and STEVENS, C.F. (1976). Bayesian Forecasting (with Discussion). J.R. Statist. Soc. B, 38, 205-47.

HUBER, P.J. (1964). Robust estimation of a location parameter. Ann.Math. Statist., 35, 73-101.

LINDLEY, D.V. and SMITH, A.F.M. (1972). Bayes estimates for the linear model (with Discussion). J.R.Statist.Soc. B, 34, 1-41.

MASRELIEZ, C.J. (1975). Approximate non-Gaussian filtering with linear state and observation relations. I.E.E.E. Trans. Aut. Control, AC-20, 107-110.

MASRELIEZ, C.J. and MARTIN, R.D. (1977). Robust Bayesian estimation for the linear model and robustifying the Kalman Filter. I.E.E.E. Trans. Aut. Control, AC-22, 361-71.

NAYLOR, J. and SMITH, A.F.M. (1981). Approximate inferences for a mixture distribution. In preparation.

O'HAGAN, A. (1979). On outlier rejection phenomena in Bayes inference. J.R. Statist. Soc. B. 41, 358-67.

PETTIT, L. and SMITH, A.F.M. (1981). Bayes methods for outliers. In preparation.

RAMSAY, J.O. and NOVICK, M.R. (1980). PLU Robust Bayesian Decision Theory: Point Estimation. J. Amer. Statist. Ass. 75, 901-07.

RELLES, D.A. and ROGERS, W.H. (1977). Statisticians are fairly robust estimators of location. J. Amer. Statist. Ass. 72, 107-11.

SMITH, A.F.M. and SPIEGELHALTER, D.J. (1981). Bayes factors and choice criteria for linear models. J.R. Statist. Soc. B, 42, 213-20.

SPIEGELHALTER, D.J. (1977). A test for normality against symmetric alternatives. Biometrika 64, 415-18.

SPIEGELHALTER, D.J. (1978). Adaptive Inference using a Finite Mixture Model. Ph.D. dissertation. University College London.

SPIEGELHALTER, D.J. (1980). An omnibus test for normality for small samples. Biometrika, 67, 493-96.

SPIEGELHALTER, D.J. (1981). Sampling properties of a finite mixture model. Unpublished manuscript. University of Nottingham.

STIGLER, S.M. (1977). Do robust estimators work with real data? Ann. Statist. 5, 1055-98.

TUKEY, J.W. (1960). A survey of sampling from contaminated distributions. In Contributions to Probability and Statistics: Essays in Honour of Harold Hotelling. Stanford University Press.

UTHOFF, V.A. (1970). An optimum test property of two well-known statistics. J. Amer. Statist. Ass. 65, 1597-1600.

UTHOFF, V.A. (1973). The most powerful scale and location invariant test of the normal against the double exponential. Ann. Statist. 1, 170-74.

WEST, M. (1981). Robust sequential approximate Bayesian estimation. J.R. Statist. Soc. B, 43. 157-66.

THE PROBABILITY INTEGRAL TRANSFORMATION FOR NON NECESSARILY ABSOLUTELY CONTINUOUS DISTRIBUTION FUNCTIONS, AND ITS APPLICATION TO GOODNESS-OF-FIT TESTS

by

J.P. Raoult, D. Criticou, D. Terzakis

(Equipe de Recherches Associée au CNRS 900
Calcul des Probabilités et Statistique
Université de Rouen*, FRANCE)

Abstract

To any probability measure Q on $\mathbb{R}^k$, it is possible to associate a probability transition (i.e. a Markov kernel) Q^R, from $\mathbb{R}^k$ to $[0,1]^k$, such that the composition of Q by Q^R is the uniform probability (or Lebesgue measure) on $[0,1]^k$; Q^R is called the probability integral transform (p.i.t) of Q. $\mathcal{P}$ being a class of probability measures on a space of observations Z, let φ be a $\mathcal{P}$ sufficient mapping from Z to a space Y, and ξ a mapping from Z to $\mathbb{R}^k$; let, for every z, $\psi(z)$ be the probability associated to point $\xi(z)$ (in $\mathbb{R}^k$) by the p.i.t of the law of ξ for any probability measure deduced from P (in $\mathcal{P}$) through conditioning by $[\varphi = \varphi(z)]$; for any P in $\mathcal{P}$, the composition of P by ψ is the uniform probability on $[0,1]^k$; this property gives a general method of construction of goodness-of-fit tests for $\mathcal{P}$; properties of these tests are discussed, with respect to the choice of the mappings φ and ξ.

* BP 67, F 76130 Mont Saint-Aignan

The revised version of this paper has been written (in March 82) during the stay of J.P. Raoult at the Institut National de la Statistique et des Etudes Economiques, Paris.

Key-words : Goodness-of-fit tests, Multivariate analysis, Regular conditional probability, Sufficiency, Probability integral transformation.

AMS 1970 subject classification : Primary 62H15, Secondary 62G10, 60A10

1. Introduction

F.J. O'Reilly and C.P. Quesenberry have proposed ([5]) a method of construction of distribution free tests of fit for a composite hypothesis $\mathcal{P}$ (class of probability measures on a measurable space $(Z,\mathcal{Z})$) under the following essential conditions, that we rephrase, from [5], into a more general frame which will be useful in the sequel of this paper :

(i) one considers a measurable mapping φ, from Z into the measurable space $(Y,\mathcal{Y})$, which is $\mathcal{P}$-sufficient ; let $\mathcal{P}^{\varphi}$ denote a regular φ-conditional probability, which is valid for every $P(\in\mathcal{P})$: for every y, we shall interpret $\mathcal{P}^{\varphi}(y)$ (which is denoted by some authors $\mathcal{P}^{[\varphi = y]}$) as "the" probability measure for observations z in Z, knowing that $\varphi(z) = y$.

(ii) one considers a measurable mapping ξ, from Z into $\mathbb{R}^k$ such that, for every P in $\mathcal{P}$, $\xi(P)$ is absolutely continuous, with respect to the Lebesgue measure $\lambda^{\otimes k}$ and, for $\varphi(P)$ - almost y, $\xi.(\mathcal{P}^{\varphi}y)$ (i.e. "the" law of ξ under condition $[\varphi=y]$) is absolutely continuous.

More precisely, in [5], Z is $\mathbb{R}^n$ $(n\geq k)$ and ξ is a projection of $\mathbb{R}^n$ onto $\mathbb{R}^k$; typically, $\xi(x_1, \ldots, x_n)=(x_1, \ldots, x_k)$ or $\xi(x_1, \ldots, x_n)=(x_n, \ldots, x_{n-k+1})$.

The central idea of O'Reilly and Quesenberry's method is to consider the multivariate "probability integral transformation" (p.i.t) of $\xi.(\mathcal{P}^{\varphi}y)$, as has been defined by Rosenblatt [8] : let Q be a probability on $\mathbb{R}^k$, absolutely continuous with respect to $\lambda^{\otimes k}$; then the p.i.t. associated to Q is the mapping Q^R (R for "Rosenblatt"), from $\mathbb{R}^k$ into $[0, 1]^k$, defined by

$$Q^R(x_1, \ldots,x_k)=(F_1(x_1), F_2^1(x_1 ; x_2), \ldots, F_k^{1,\ldots,k-1}(x_1,\ldots,x_{k-1};x_k))$$

where F_1 is the d.f. of the first projection of Q, and, for $j(2\leq j\leq k)$, $F_j^{1,\ldots,j-1}(x_1,\ldots,x_{j-1}\ ;\ .)$ is the d.f. of a version of the conditional probability on the jth component, with respect to the values $(x_1,\ldots,x_{j-1})$ of the first j-1 components. It is established in [8] (and is, in fact, straightforward) that $Q^R.Q = U^{\otimes k}$ (U denoting now, and in all the sequel, the uniform probability on [0, 1]).

From this, O'Reilly and Quesenberry establish that, if we denote by G_y the p.i.t. associated to $\xi.(\mathcal{P}^{\varphi}y)$, the law, with respect to every $P(\in\mathcal{P})$, of the random vector $z \leadsto G_{\varphi(z)}(\xi(z))$ is $U^{\otimes k}$; so, the problem of testing (on Z) is transformed into the simple testing problem of uniformity (on $[0, 1]^k$). O'Reilly and Quesenberry also give in their paper a multivariate generalization (with Z equal to $(\mathbb{R}^p)^n$ and $(\mathbb{R}^p)^k$ instead of $\mathbb{R}^p$) of the approach that we have just described, as well as the computation of $G_{\varphi(z)}(\xi(z))$ in some cases of practical importance (e.g. for tests of normality).

It is interesting to remark that, if φ is $\mathcal{P}$- complete, the mapping $z \leadsto \xi.(\mathcal{P}^{\varphi}(\varphi(z)))$ gives the unique minimum variance unbiaised (U.M.V.U.) estimator of $\xi.P$ (for $P \in \mathcal{P}$) (unbiasedness meaning here that, for every Borel subset A of $\mathbb{R}^k$,

$$\int [\xi.(\mathcal{P}^{\varphi}(\varphi(z))](A)\ dP(z) = P[\xi\epsilon A])$$

and we know from [9] that the cases considered by O'Reilly and Quesenberry are exactly those where there exists an unbiased estimator of the density of the law of ξ, this estimator being the density of $\xi.(\mathcal{P}^{\varphi}(\varphi(z))$ (here estimator is of course considered in $L_1(\mathbb{R}^k, \lambda^{\otimes k})$).

Our aim in this paper is to present results analogous to these of O'Reilly and Quesenberry, without any assumption of absolute continuity ; it would be so, for instance, if all distributions in $\mathcal{P}$ were discrete ; it is classical, in such cases, to increase the power of tests by the use of randomization ; so, the tests that we propose are random tests and, in their construction, we use a generalization of the p.i.t. (defined as a probability transition instead of a function) which has partially been introduced, for instance, in [4] ; it follows, in particular, that, in the most common cases ($Z=\mathbb{R}^n$, i.i.d. observations according to $\tilde{P}$($\tilde{P} \in \tilde{\mathcal{P}}$, a class of absolutely continuous probabilities on $\mathbb{R}$), $X = \mathbb{R}^m$) we are able to propose distribution-free goodness-of fit tests for $\tilde{\mathcal{P}}$ even in some cases where there does not exist any unbiaised estimator of density of $\tilde{P}$ (inside $\tilde{\mathcal{P}}$).

Or course, the fact that we do not impose $\xi.(\mathcal{P}^{\varphi}y)$ to be almost surely absolutely continuous provides a great flexibility in the choice of φ and ξ ; we therefore conclude with some remarks on the influence on power and robustness of these two mappings.

2. Multivariate Probability Integral Transformations (p.i.t.)

a) Univariate p.i.t.

Let Q be a probability measure on $\mathbb{R}$, and let F be its d.f. ; we note Q^R the probability transition, from $\mathbb{R}$ to [0, 1], which is defined by :

(i) if x is a continuity point of F(i.e. $Q(\{x\})=0$). $Q^R(x)$ (or, briefly, $Q^R x$)is the Dirac measure at point F(x),

(ii) if x is a discontinuity point of F, $Q^R x$ is the uniform probability measure on the "jump interval" [F(x-), F(x)].

It is well known that $Q^R.Q = U$ ($Q^R.Q$ is the probability measure defined on [0, 1] by

$$[Q^R.Q](B) = \int (Q^R x)(B) \, dQ(x) \quad) ;$$

this is evidently a generalization of relation $Q^R.Q = U$ given in the introduction in the case of Q absolutely continuous (or even just F continuous, in which case $(Q^R x)(B) = 1$ if $F(x) \varepsilon B$ and 0 if not).

An other way to look at this, more familiar, perhaps, to applied statisticians, is to consider the mapping G from $\mathbb{R} \times [0, 1]$ to $[0, 1]$ defined by

(i) if x is a continuity point of F, $G(x,y) = F(x)$ for all y,

(ii) if x is a discontinuity point of F, $G(x,y) = F(x-) + \frac{y}{F(x)-F(x-)}$,

and to remark that $G(P \otimes U) = U$.

b) <u>Simple multivariate p.i.t.</u>

Let Q be a probability measure on $\mathbb{R}^k$; let us fix, for each $j(2 \leq j \leq k)$ a regular conditional probability $Q_j^{1,\ldots,j-1}$, for the jth component with respect to the first $j-1$ components ; following the notations of a, we can consider, for each $j(2 \leq j \leq k)$ and each $x_1, \ldots, x_{j-1}, x_j$, the probability $[Q_j^{1,\ldots,j-1}(x_1,\ldots,x_{j-1})]^R(x_j)$ on $[0, 1]$. Then we define

$$Q^R(x_1,\ldots,x_k) = \bigotimes_{j=1}^{k} \{[Q_j^{1,\ldots,j-1}(x_1,\ldots,x_{j-1})]^R(x_j)\}$$

with the convention that the first factor ($j=1$) in r.h.s. is $Q_1^R(x_1)$, with Q_1 the first margin of Q.

It is quite elementary that we still have $Q^R.Q = U^{\otimes k}$; we just sketch the computation for $k=2$:

$$[Q^R.Q]([0,a] \times [0,b])$$

$$= \int [Q^R(x_1, x_2)]([0,a] \times [0,b]) \, dQ(x_1, x_2)$$

$$= \int \left[(Q^R x_1)[0,a]\right]\left[((Q_2^1 x_1)^R x_2)[0,b]\right] dQ(x_1, x_2)$$

$$= \int \left[(Q_1^R x_1)[o,a]\right] \{\int ((Q_2^1 x_1)^R x_2)[0,b] \, d(Q_2^1 x_1)(x_2)\} dQ_1(x_1)$$

$$= \int \left[(Q^R x_1)[o,a]\right] b \, dQ_1(x_1) = a\, b$$

c) Multiple multivariate p.i.t.

Let $X = X_1 x \ldots x X_m$ and, for each $j (1 \leq j \leq m)$, let η_j be a measurable mapping from X_j into $\mathbb{R}^{k_j}$; let $k = \sum_j k_j$.

Let Q be probability measure on X ; let us suppose that, for each $j(2 \leq j \leq m)$, there exists a regular conditional probability $Q_j^{1,\ldots,j-1}$ on X_j, with respect to points in $X_1 x \ldots x X_{j-1}$; let Q_1 be the projection of Q on X_1 ; than we define

$$Q^R(x_1,\ldots,x_m) = (\eta_1 . Q_1)^R(\eta_1(x_1)) \blacksquare (\eta_2 . (Q_2^1(x_1)))^R(\eta_2(x_2)) \blacksquare \ldots \blacksquare (\eta_m . (Q_m^{1,\ldots,m-1}(x_1,\ldots,x_{m-1})))^R(\eta_m(x_m))$$

(all p.i.t. in r.h.s. being defined as in b)

We still have $Q^R . Q = U^{\blacksquare k}$ (the proof goes along the same lines as in b).

We remark that b can be seen as a particular case of c in two ways : either we take $m=1$, $X=\mathbb{R}^k$, η the identity mapping, or we take $m=k$, $X_1 = \ldots = X_k = \mathbb{R}$, η_j the jth projection of $\mathbb{R}^k$ on $\mathbb{R}$.

3. Principle of tests

a) Let $\mathcal{P}$ be a class of probability measures on the measurable space $(Z,\mathcal{Z})$: let φ be a measurable mapping, with values in $(Y,\mathcal{Y})$, which is $\mathcal{P}$-sufficient, with regularity of conditioning for each P.

Let P^{φ} be a transition probability, from Y to Z, which is, for each P $(\in\mathcal{P})$, a regular φ-conditional probability.

Let $\xi (=(\xi_1,\ldots,\xi_k))$ be a k-dimensional random vector.

We assume that $\mathcal{P}^{\varphi}$ is ξ-regular in the sense that we can ensure the measurability of all mappings (for $2\leq j\leq k$)

$$(y,x_1,\ldots,x_j) \rightsquigarrow \{[\xi.(\mathcal{P}^{\varphi}y)]_j^{1,\ldots,j-1}(x_1,\ldots,x_{j-1})\}(]-\infty,x_j])$$

(value, at x_j, of "the" d.f of ξ_j, conditionaly to $\varphi(z)=y$, $\xi_1(z) = x_1,\ldots,$ $\xi_{j-1}(z) = x_{j-1}$). It follows from the method of construction of regular conditional probabilities (see [6]) that $\mathcal{P}^{\varphi}$ can be chosen ξ-regular if $(Z,\mathcal{Z})$ and $(Y,\mathcal{Y})$ are separable standard Borel spaces, and φ is a map of Z onto Y (conditions which are evidently satisfied in any realistic case !)

b) Theorem

> Let $\psi(z) = [\xi.(\mathcal{P}^{\varphi}(\varphi(z))]^R(\xi(z))$.
> Then, for every $P \in \mathcal{P}$, $\psi.P = U^{\otimes k}$

Proofs.

We first remark that, by the definition of p.i.t., for every z, $\psi(z)$ is a probability measure on $[0, 1]^k$ - It follows from the hypothesis of ξ-regularity of $\mathcal{P}^{\varphi}$ that, for every measurable subset A of $[0, 1]^k$, the map $z \rightsquigarrow [\psi(z)](A)$ is measurable, so that, for any probability measure P on Z, $[\psi.P](A)$ $(=\int_Z(\psi(z))(A)\,dP(z))$ makes sense.

Let now consider $P(\in\mathcal{P})$, and let us note P^{φ} instead of $\mathcal{P}^{\varphi}$ (insisting on the fact that this is a regular φ-conditional probability for P) ; writing $P' = \varphi.P$ and keeping in mind that $\psi(z)$ is a probability measure on [0, 1], we get

$$\int_Z[\psi(z)](A)\ dP(z)$$

$$= \int_Y\{\int_Z\big[[\xi.(P^{\varphi}\varphi(z))]^R(\xi(z))\big](A)\ d(P^{\varphi}y)(z)\}\ dP'(y)$$

$$= \int_Y\{\int_Z[[\xi.(P^{\varphi}y)]^R(\xi(z))](A)\ d(P^{\varphi}y)(z)\}\ dP'(y)$$

$$= \int_Y\{\int_{\mathbb{R}^k}\big[[\xi.(P^{\varphi}y)]^R(x)\big](A)\ d(\xi.(P^{\varphi}y))(x)\}\ dP'(y)$$

$$= \int_Y U^{\otimes k}(A)\ dP'(y) = U^{\otimes k}(A)$$

∎

For the sake of simplicity, we have presented this theorem in the case of simple multivariate p.i.t. (see 2 b) ; but it obviously remains valid in the case of multiple multivariate p.i.t. (see 2 c) : nothing has to be changed neither in the theorem nor in its proof ; the only technical modification is in the condition of ξ - regularity : we now have $X = X_1 \times \ldots \times X_m$ and, for each j, $\eta_j : X_j \to \mathbb{R}^{k_j}$; we postulate, for each $j (1 \leq j \leq m)$ and each $i (1 \leq i \leq k_j)$, the measurability of the mapping

$$(y, x_1, \ldots, x_{j-1}, u_1, \ldots, u_{i-1}, u_i) \rightsquigarrow$$

$$\{ \left[[\eta_j . P_j^{1,\ldots,j-1}](x_1,\ldots,x_{j-1}) \right]_i^{1,\ldots,i-1} (u_1 \ldots, u_{i-1}) \} (]-\infty, u_i])$$

from $Y \times X_1 \times \ldots \times X_{j-1} \times \mathbb{R}^i$ into $[0,1]$

(with obvious conventions for $j = 1$ or $i = 1$) ; this condition is, once again, satisfied if, in addition to conditions on Z, Y and φ given previously, all spaces X_j are also separable standard Borel spaces.

c) The principle of the goodness-of-fit test for family $\mathcal{P}$ is now clear: having once chosen φ and ξ, and observed z, we compute the probability $\psi(z) (= [\xi . \mathcal{P}^{\varphi}(\varphi(z))]^R (\xi(z))$ and draw a point u at random in $[0,1]^k$ according to probability $\psi(z)$; this draw will be made through a sequence of k operations which are either deterministic (when we are at a continuity point of the d.f which is considered at this stage) or according to the uniform probability (see the comments at the end of 2 a, which we apply recursively). Then we apply to u a test of uniformity on $[0,1]^k$.

Let us now give some details on the computation of $\psi(z)$ in one of the most common cases : $Z = (\mathbb{R}^p)^n$, $X = (\mathbb{R}^p)^m$ (so that $k = mp$) and $\xi(x_1,\ldots,x_n) = (x_{\alpha(1)},\ldots,x_{\alpha(m)})$, with the $\alpha(j)$ $(1 \leq j \leq m)$ all distinct (usually, one has either $(\alpha(1),\ldots,\alpha(m)) = (1,\ldots,m)$, or $(\alpha(1),\ldots,\alpha(m)) = (n,\ldots,n-m+1))$.

Then, denoting as usual by upper indices the conditionings, and by lower indices the projections (made <u>after</u> the conditionings) we get, by definition of multivariate p.i.t,

$$[\xi.(\mathcal{P}^{\varphi}y)]^{R}(t_1,\dots,t_k) = \bigotimes_{j=1}^{m} [[\mathcal{P}^{\varphi}y]_{\alpha(j)}^{\alpha(1),\dots,\alpha(j-1)}(t_1,\dots,t_{j-1})]^{R}(t_j)$$

(with the first factor of r.h.s ($j = 1$) equal to $[[\mathcal{P}^{\varphi}y]_{\alpha(1)}]^{R}(t_1))$, which is, P-almost surely for any P in $\mathcal{P}$, equal to

$$\bigotimes_{j=1}^{m} [[\mathcal{P}_{\alpha(j)}^{\varphi,\alpha(1),\dots,\alpha(j-1)}](y,t_1,\dots,t_{j-1})]^{R}(t_j) \quad ;$$

it follows that, for $z = (x_1,\dots,x_n)$, we can note

$$\psi(z) = \bigotimes_{j=1}^{m} [\mathcal{P}_{\alpha(j)}^{\varphi,\alpha(1),\dots,\alpha(j-1)}(\varphi(z),x_{\alpha(1)},\dots,x_{\alpha(j-1)}]^{R}(x_{\alpha(j)}) \quad ,$$

with $[\mathcal{P}_{\alpha(1)}^{\varphi}(\varphi(z))]^{R}(x_{\alpha(1)})$ as first factor.

Let us now particularize to the case where, for every $n (\geq 1)$ we want to test whether $\tilde{P}$ (probability on $\mathbb{R}^p$) belongs to a class $\tilde{\mathcal{P}}$ by observing the realization of n i.i.d. random vectors with law $\tilde{P}$, and by using a sufficient statistic φ_n (on $Z = (\mathbb{R}^p)^n$) ; φ_n is sufficient for the class $\mathcal{P}_n = \{\tilde{P}^n ; \tilde{P} \in \tilde{\mathcal{P}}\})$. We recall (see [5]) that the sequence $(\varphi_n)_{n\geq 1}$ is said to be doubly transitive if, for each n, the mappings

$$(x_1,\dots,x_n) \rightsquigarrow (\varphi_n(x_1,\dots,x_n) \ , \ x_n)$$

and

$$(x_1,\dots,x_n) \rightsquigarrow (\varphi_{n-1}(x_1,\dots,x_{n-1}),x_n)$$

are measurably equivalent (i.e. induce the same σ-algebra on $(\mathbb{R}^p)^n$) ; let us choose $\xi_n(x_1,\dots,x_n) = (x_n,\dots,x_{n-m})$; then it is elementary to verify (and details can be found in [1]) that (exactly as in the absolutely continuous case) the computation of $\psi(x_1,\ldots,x_n)$ simplifies into

$$\bigotimes_{j=0}^{m-1} [(\tilde{\mathcal{P}}^{n-j})_{n-j}^{\varphi_{n-j}}(\varphi_{n-j}(x_1,\ldots,x_{n-j}))]^R(x_{n-j})$$

(for general considerations on transitivity in a Bayesian framework, see also [2] and [3]).

All these considerations on computation (and in particular the notion of double transitivity, and the simplifications that it provides) remain evidently valid if, instead of $\mathbb{R}^p$, we have an abstract space T, and a mapping from T to some $\mathbb{R}^p$ (see the definition of multiple multivariate p.i.t.).

4. <u>Remarks</u>

As we had noticed in the introduction, the great flexibility that we have on φ and ξ suggests some comments on the influence of the choice of these two mappings on the qualities of the test.

Of course, it is preferable to take φ such that it eliminates as much as possible of the information which is irrelevant to the problem: does P belong or not to $\mathcal{P}$? ; in other words, it would be interesting to take φ minimal sufficient. Such a choice minimizes the "risk" that φ could also be sufficient for a class $\mathcal{P}'$ greater than $\mathcal{P}$; in such a case, elements of $\mathcal{P}'-\mathcal{P}$ would be, with our method, indistinguishable from those of $\mathcal{P}$. Of course, the worst choice for φ would be to take the identity map of Z ; then $\xi.(\mathcal{P}^{\varphi}z)$ would be the Dirac measure at point $\xi(z)$, and $\psi(z)$ would be $U^{\otimes k}$, and remain so whatever the probability P would be !

As to ξ, our interest is of course to avoid as much as possible the risk of confounding $\xi.(\mathcal{P}^{\varphi}y)$ with $\xi.(P'^{\varphi}y)$, for P' not in $\mathcal{P}$; from this point of view, an ideal choice would be, if $Z = \mathbb{R}^N$, to use as ξ the identity map of $\mathbb{R}^N$, but the difficulty of computation of $[\xi.(\mathcal{P}^{\varphi}y)]^R$ (with ξ a k-dimensional random vector) may increase rapidly with k , and then we have to compromise.

A comparaison with O'Reilly and Quesenberry's paper is interesting at this stage. Limiting our comments, for the sake of simplicity, to the univariate part of their paper, let us recall that they only consider mappings ξ which are projections from $\mathbb{R}^n$ into $\mathbb{R}^k$, and that they suppose that $P = \tilde{P}^{\otimes n}$ ($\tilde{P} \in \tilde{\mathcal{P}}$) and φ is symmetric; as they insist on having $\xi.(\mathcal{P}^{\varphi}y)$ absolutely continuous for almost every y , they are led, for the sake of loosing as little information as possible by the use of ξ , to introduce the absolute continuity rank (a.c.r) of $\tilde{\mathcal{P}}$ with respect to φ , that is to say the maximum value of h ($1 \leq h \leq n$) such that $\xi_h.(\mathcal{P}^{\varphi}y)$ is a.s. absolutely continuous (with $\xi_h(x_1,\dots,x_n) = (x_1,\dots,x_n)$); let us remark that, φ being symmetric, the ordering of coordinates is here irrelevant. They remark that, "for many families", the a.c.r. is of the form n-c, with c a fixed integer (typically, the dimension of the space Y) ; actually, for a great many of these families, $\mathcal{P}^{\varphi}y$ is concentrated on a (n-c) - dimensional manifold in $\mathbb{R}^n$, say M_y , such that, for each $(x_1,\dots,x_{n-c})$ there is at most one point un M_y with $(x_1,\dots,x_{n-c})$ as its n-c first coordinates ; then when we construct "our" p.i.t., with ξ the identify map of $\mathbb{R}^n$, we get $[\mathcal{P}^{\varphi}(\varphi(z))]^R(z)$ as the product of n factors, the first n-c among them being Dirac probability measures (they exactly correspond to the "classical" p.i.t. used in such a case by O'Reilly and Quesenberry), and the last c among them being the uniform measure on [0,1] (that is to say absolutely non informative on $\tilde{P}$) ; in such a case, our proposal is no amelioration compared to O'Reilly and Quesenberry's.

But, even when $\mathcal{P}^{\varphi}y$ is concentrated on a (n-c) - dimensional manifold M_y , it can happen that the intersection of M_y with the cylinder of basis $\{(x_1,\dots,x_{n-c})\}$ is not reduced to one point, with possibly not the same conditional probability (concentrated on this intersection) for all values of y ; then the randomized transformation that we propose in this paper does indeed preserve more information than O'Reilly and Quesenberry's non randomized transformation, and there is some hope that our tests could be more powerful than theirs.

Of course, there are cases where O'Reilly and Quesenberry's theory is not applicable at all, because there is no projection ξ_k such that $\xi_k.(\mathcal{P}^{\varphi}y)$ would be absolutely continuous (in such a case, they say that the a.c.r. is zero). Such a case is obtained, for instance, if one knows a given interval $]u,v[$ (with $-\infty \leq u < v \leq +\infty$) and a positive valued function f on $]u,v[$, such that, for each a $(u < a < v)$ $\int_u^a f(t)\,dt < \infty$; than we define $\tilde{\mathcal{P}}$, on $\mathbb{R}$, as the family of all probabilities $\tilde{P}_a$ with density $f_a = 1_{]u,a[}\, f/\int_o^a f(t)\,dt$ (a is unknown, and we proceed to n independent observations according to probability $\tilde{P}_a$) . The mapping $\varphi_n((x_1,\ldots,x_n) \rightsquigarrow \sup(x_1,\ldots,x_n))$ is sufficient and complete ; let us notice that this is a case (see [9]) where there is no unbiaised estimator of f_a , because the unique unbiaised estimator of P_a based on φ_n provides an estimation whose d.f has a jump (of magnitude $\frac{1}{n}$) at point $\sup(x_1,\ldots,x_n)$. The computation of $[\xi_k.(\mathcal{P}^{\varphi}(\varphi_n(x_1,\ldots,x_n))]^R(x_1,\ldots,x_k)$ requires the consideration of different cases according to the position of the different components of $(x_1,\ldots,x_k)$ relatively to $\sup(x_1,\ldots,x_n)$; the smaller k is, the easier this computation turns out to be (let us remark that this is a case where there is no double transitivity : it is clear that the knowledge of couple (x_1,x_2) is strictly more informative than that of $(\sup(x_1,x_2),x_2)$; that explains the intricacy of computations).

The power of the tests which have been constructed relies on the choice of ξ (typically, as in the example above, on the choice of dimension k) , and on the choice of the test for uniformity which has to be used <u>in fine</u> ; on this last point, a bibliographical survey (including in particular the very rich work [7]) can be found in [10] ; empirically, it would be interesting, in the example above, to consider, for different values of k, and some important modifications f^* of function f , the behavior (for n great enough) of $\psi.(\tilde{P}_a^{*\otimes n})$ (where ψ is (see 3a) adapted to the class $\mathcal{P}$ $(=\{\tilde{P}^{\otimes n}$, $\tilde{P} \in \tilde{\mathcal{P}}\}))$ and then, choosing, with the help of [10], a well adapted uniformity test, to compute its power at "point" $\psi.(\tilde{P}_a^{*\otimes n})$.

The study of robustness would imply the computation, starting from the (small) distance (in variation, or Hellinger) between P^* and $\mathcal{P}$, of the distance between $\psi(P_a^{*\otimes n})$ and $U^{\otimes k}$. A first empirical step could consist in explicit computations for some minor changes of function f, or for the presence of outliers: f being unchanged, introduction of a small proportion of values of parameter a different from the "common" value.

References

[1] CRITICOU (D.) (1981) - Tests de normalité multidimensionnelle (méthode générale, usage de transformations de Rosenblatt). Documents de travail, Equipe de calcul des Probabilités et Statistique, Université de Rouen (France) 1981-1.

[2] FLORENS (J.P.), MOUCHART (M.) et ROLIN (J.M.) (1980). Réductions dans les expériences bayesiennes séquentielles Coll. Processus aléatoires et problèmes de prévision, Cahiers C.E.R.O. (Bruxelles), 22, 3-4, 353-362.

[3] FLORENS (J.P.) et MOUCHART (M.) (1980) - Initial and sequential reduction of Bayesian experiments, Discussion Paper 8015, C.O.R.E., Université Catholique de Louvain (Belgium)

[4] MOORE (D.S.) and SPRUILL (M.C) (1975) - Unified large sample theory of general chi-square statistics for goodness-of-fit, Ann. Stat., 3,3, 599-616

[5] O'REILLY (F.J.) and QUESENBERRY (C.P.) (1973). The conditional probability integral transformation and applications to obtain composite chi-square goodness-of-fit tests. Ann. Stat., 1,1,74-83.

[6] PARTHASARATY (K.R.). Probability measures on metric spaces, Academic Press, New York.

[7] QUESENBERRY (C.P.) and MILLER (F.L, Jr) (1977). Power studies of some tests for uniformity, J. Statist. Comput. Simul., 5 169-191.

[8] ROSENBLATT (M.) (1952) - Remarks on a multivariate transformation, Ann. Stat., 23, 470-472.

[9] SEHEULT (A.H.) and QUESENBERRY (C.P.) (1971). On unbiaised estimation of density functions. Ann. Math. Stat. 42, 1434-1438.

[10] TERZAKIS (D.) (1981). Tests de normalité multidimensionnelle (méthode par transformation orthogonale. Calculs de puissance). Documents de travail, Equipe de calcul des Probabilités et Statistique, Université de Rouen (France), 1981-2.

SIMULATION IN THE GENERAL FIRST ORDER AUTOREGRESSIVE PROCESS (UNIDIMENSIONAL NORMAL CASE)

by

Paul Doukhan*

Abstract

The main object of this work is to show that some theoretical results concerning autoregressive processes [4] are indeed applicable : first, we choose discretization parameters for the computation of non parametric kernel estimators for these processes; then, we investigate some "bad" cases and some "good" cases; it seems that effective computations generally give better results than those obtained in theory, finally we study the relation between the deterministic case of iterations and the nondeterministic case of autoregressive process. In addition, we describe the behaviour of the invariant measures associated with the relevent process when there is little white noise.

Key-words : Autoregressive process, Nonparametric kernel estimators, Invariant measures, Simulation

AMS/MOS : Primary 62M10, Secondary, 62E25, 62G05

*Presently at the University of Rouen, Equipe de Recherche de Calcul des Probabilités et Statistiques (E.R.A. 900), 76130 Mont Saint Aignan, France. This work was supported by the biometry laboratory of I.N.R.A., Jouy en Josas 78, France.

1. GENERALITIES

1.1. Introduction

We study here the practical validity of theoretical results by M. Ghindès and P. Doukhan concerning the estimation of the regression function of the Markov process $(X_n) : X_{n+1} = f(X_n) + \varepsilon_n$ where (ε_n) is a white noise process. Let us recall that such processes are quite frequent in practice. In fact, the evolution of discases or iterative calculations with a computer can be easily modelled by such processes.

1.2. Theory

THEOREM A [1]

If f *is a bounded measurable real valued function and* (ε_n) *an i.i.d.*-$N(0,\sigma^2)$ *sequence, the process* (X_n) *defined by* $X_{n+1} = f(X_n) + \varepsilon_n$ *is an aperiodic irreductible and geometrically ergodic Markov process.*

The density Π of the limit law of the process (X_n) can be estimated by $\hat{\Pi}_n(x) = \frac{1}{n\beta_n} \sum_{k=0}^{n-1} u\left(\frac{X_k - x}{\beta_n}\right)$ where u is the kernel defined by $u(x) = \frac{1}{2} e^{-|x|}$. The regression function f is estimated on any interval $[-a,a]$ by $\hat{f}_n(x) = \frac{\hat{B}_n(x)}{\mathrm{Max}(b,\hat{\Pi}_n(x))}$ where b and $\hat{B}_n$ are defined by : $b \leq \frac{1}{2\sqrt{2\Pi}\,\sigma} \exp\left(-(a + \|f\|_\infty)^2 / 2\sigma^2\right)$ and $\hat{B}_n(x) = \frac{1}{n\beta_n} \sum_{k=0}^{n-1} X_{k+1}\, u\left(\frac{X_k - x}{\beta_n}\right)$.

THEOREM B [2]

If f *is derivable in the sense of distributions and* $\|f\|_\infty + \|f'\|_2 < +\infty$, *there are constants* C_π *and* C_f *such that for any initial law* ν, *any real number* $a > 0$ *and any integer* n *:*

$$\int E_\nu(\hat{\Pi}_n(x) - \Pi(x))^2 \, dx < C_\pi \, n^{-2/3} ,$$

$$R_n = \int_{[-a,a]} E_\nu(\hat{f}_n(x) - f(x))^2 \, dx < \frac{C_f}{b^2} \, n^{-2/3}$$

where $\beta_n = n^{-1/3}$.

Remarks

. The kernel u chosen above also gives us uniform quadratic convergence of the estimates; the bias of our estimator converges to zero if the condition $\|f'\|_2 < \infty$ of Theorem B is replaced by $\|f'\|_1 < +\infty$.

. The expression of $\hat{f}_n$ is simplified by taking $b = 0$; this is equivalent to assume that the theoretical $\hat{f}_n$ makes use of a value b which is lower than most values of $\hat{\Pi}_n$ encountered in the simulation (this is reasonable, for n large enough, because the probability that $\hat{\Pi}_n$ takes values lower than b is $o(n^{-2/3})$; see [3]).

Let (a_n) an increasing sequence $a_n \uparrow \infty$. We estimate σ^2 by

$$\frac{1}{p}\sum_{k=K+1}^{K+p+1}\left(X_{k+1} - \hat{f}(X_k)\right)^2$$

in the following cases :

(i) $\hat{f} = \hat{f}_{n-p-p_0}$, $K = n-p$, $a = a_{n-p-p_0}$; the estimate is denoted by $\hat{\sigma}_n^2$.

(ii) $\hat{f}$ is the estimate computed on the shifted data $(X_{p+1}, \ldots, X_n)$, with $a = a_n$ and $K = 0$; the estimate is then denoted by $\tilde{\sigma}_n^2$.

THEOREM C [3]

The rates of convergence of the estimates of σ^2 *can be computed as follows. Let*

(i) $R = (\|f\|_\infty + \|\hat{f}_m\|_\infty)^2 \; P(|\varepsilon_1| > a_m - 2\|f\|_\infty + 2\alpha^{p_0}) + \rho_m$

$\alpha < 1$ *is the geometric ergodicity constant (see Theorem A) and* $m = n-p_0-p$.

(ii) $R' = (\|f\|_\infty + \|\hat{f}_{n-p}\|_\infty) \; P(|\varepsilon_1| > a_{n-p} - 2\|f\|_\infty) + \rho_{n-p}$.

(iii) ρ_k *be the product of the upper bound of the density of* (ε_1) *with*

$$\int_{[-a_k,a_k]} E_\nu\,(\hat{f}_k - f)^2 \, d\lambda.$$

Then

$$E|\hat{\sigma}_n^2 - \sigma^2| < \sigma^2\sqrt{3/p} + 2\sigma\sqrt{R} + R; \quad E|\tilde{\sigma}_n^2 - \sigma^2| < \sigma^2\sqrt{3/p} + 2\sigma\sqrt{R'} + R'.$$

Remarks

The rates of convergence seem optimal if $n \sim f^{3/2}$ and $k'\,k^{p_0} \sim n^{-2/3}$, in this case they are $0(n^{-1/3})$.

1.3. Discretization parameters for the simulations

Because simulations are not made in an asymptotic sense, we shall only estimate f on $[-1,1]$ (the general case follows by linear transforms). To compute the risk, we need two discretization parameters : j_1 is the number of points used to compute integrals on the interval $[-1,1]$, k_1 is the number of realizations of $X_1, \ldots, X_n$ used to compute expectations. In the following, the rate of convergence is computed with those data, and the approximation of "$\int_{[-1,1]} E^{X_0=0}\,(\hat{f}_n(x) - f(x))^2\,dx$" will be written R_n.

Table 1 shows the behaviour of R_n component for different values of j_1 when $k_1 = 20$, $n = 1000$, $\sigma = 0.3$ and $f(x) = 4(2x-1)\,/\,(x^2+1)^2$ which satisfies the hypothesis of Theorem B. The value $j_1 = 25$ will be fixed in the sequel.

j_1	10	25	60	110	160
R_n	0.32830	0.3250	0.32525	0.32531	0.32531

Table 1

Table 2 shows the behaviour of R_n computed with different values of k_1 when $n = 1000$, $\sigma = 0.1$ and $f(x) = x / (1 + |x|^3)$ which also satisfies the hypothesis of Theorem B. The values $j_1 = 25$ and $k_1 = 20$ will be chosen for this work because of problems due to the duration of computation.

k_1	10	20	30	40	50	60	70	80
R_n	0.0187	0.0175	0.0174	0.0168	0.0166	0.0171	0.0171	0.0172

Table 2

1.4. Graphical representations

Each drawing displays three curves :

- the continuous line represents the function f
- the dotted line represents the function $x \to E^{X_0=0}\, \hat{f}_n(x)$
- the interrupted line represents the function $x \to \hat{f}_n(x)$ (a realization).

For instance, Figure 1 is the case $\sigma^2 = 0.35$; $n = 10, 100, 1000, 10000$.

$$f(x) = \begin{cases} e^{-2x^2}, & x \geq 0 \\ (1 + 4x) / (1 + x^2), & x < 0 \end{cases}$$

2. STUDY OF SOME BAD CASES

2.1. A function which satisfies the theoretical hypothesis

$$f(x) = \begin{cases} 1-4x^2, & |x| \leq 1/2 \\ 0, & |x| > 1/2 \end{cases}$$

In the case $\sigma^2 = 0.1$ the estimation is better than in the case $\sigma^2 = 0.01$ as it is shown in Figure 2. The phenomenon results from the fact that the observations are almost never negative if $\sigma^2 = 0.01$, so that f is not well estimated for $x < 0$.

n	10	100	1000	10000
R_{M}	1.31842	0.13555	0.01820	0.00313

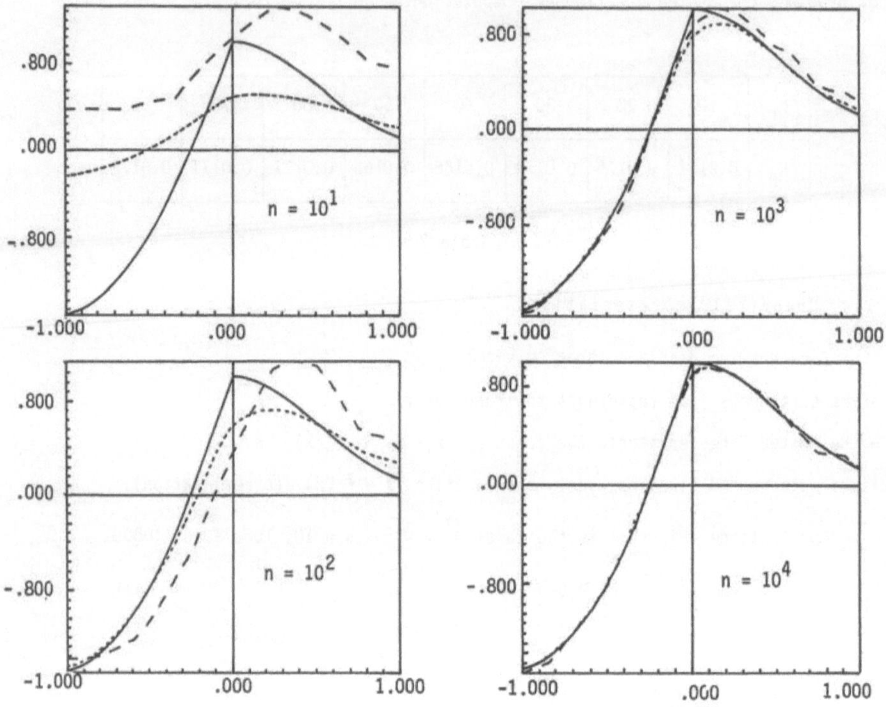

Figure 1

n		0	100	1000	10000
R [illegible]	$\sigma^2=.1$	.38955	.13842	.02319	.00405
	$\sigma^2=.01$	.54125	.55750	.45268	.08970

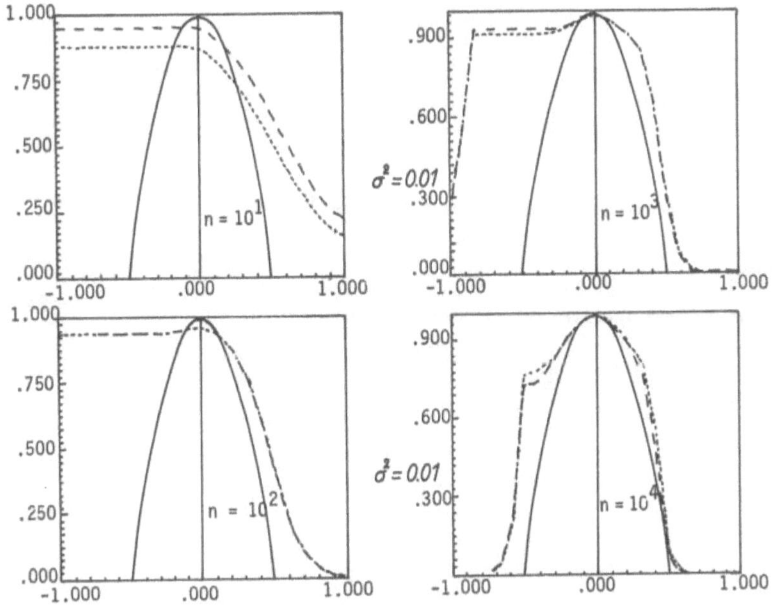

Figure 2

2.2. Some functions which do not satisfy the theoretical hypothesis

2.2.1. Discountinuous functions

$$f(x) = \begin{cases} 2x + 1 , & x < 0 \\ 2x - 1 , & x \geq 0 \end{cases} \qquad \sigma^2 = 1;\ n = 1000;\ R_n = 0.907 \text{ (Figure 3-a)}$$

$$f(x) = \begin{cases} 0 , & x < 0 \\ 1 , & x \geq 0 \end{cases} \qquad \sigma^2 = 0.1;\ n = 1000;\ R_n = 0.23 \text{ (Figure 3-b)}$$

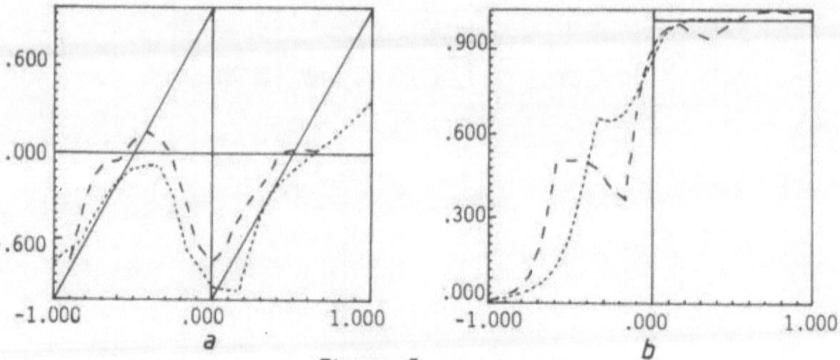

Figure 3

Case a is better than case b because, in this case, the range of f is the whole line so that the samples (X_n) are distributed over the whole line.

2.2.2. A rapidly varying function

$$f(x) = \sin (1000x);\ \sigma^2 = 0.1;\ n = 1000;\ R_n = 1.01 \quad \text{(Figure 4)}.$$

The regression function is very badly estimated because the function adds some noise in the process. Such a phenomenon is better modelled as a sequence of i.i.d. variables.

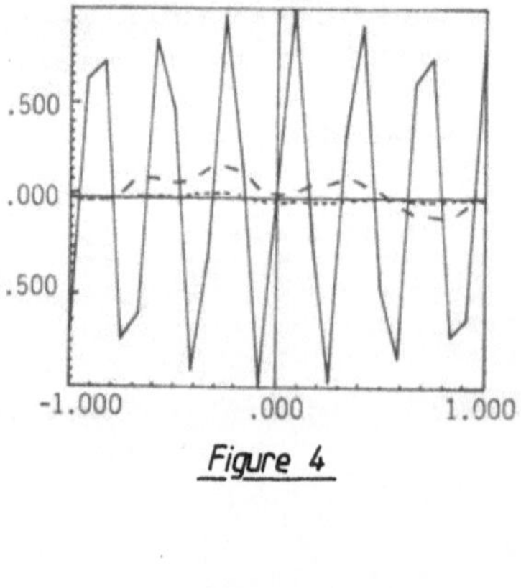

Figure 4

3. ESTIMATION OF THE VARIANCE OF THE WHITE NOISE

3.1. Comparing the methods

Here we take the function f already used in 1.3 (Figure 1) and $n = 1000$. We choose $p = 50$ and $n = 1000$ to optimize risk and computing time.

σ^2	$E\tilde{\sigma}_n^2$	$E\lvert\tilde{\sigma}_n^2-\sigma^2\rvert$	R_n
.10000	.52401	.43399	.06245
.20000	.28214	.08256	.01358
.30000	.32968	.06735	.01864
.40000	.41361	.07240	.01874
.50000	.56189	.06919	.02152
.60000	.60268	.11421	.02131
.70000	.74964	.10843	.02242
.80000	.93106	.12827	.02486
.90000	.90463	.11179	.02990
1.00000	1.06540	.13195	.02533

PO = 0

σ^2	$E\hat{\sigma}_n^2$	$E\lvert\hat{\sigma}_n^2-\sigma^2\rvert$	R_n
.10000	.11542	.02283	.06245
.20000	.22704	.02720	.01358
.30000	.30938	.05980	.01864
.40000	.43966	.07760	.01874
.50000	.59386	.10935	.02152
.60000	.68268	.11705	.02131
.70000	.75130	.11019	.02242
.80000	.86879	.09284	.02486
.90000	.89342	.13945	.02990
1.00000	1.18685	.15629	.02533

PO = 1

σ^2	$E\hat{\sigma}_n^2$	$E\lvert\hat{\sigma}_n^2-\sigma^2\rvert$	R_n
.10000	.11087	.01858	.05579
.20000	.22092	.03732	.01261
.30000	.34136	.07581	.01671
.40000	.40214	.07278	.01977
.50000	.55642	.07592	.02238
.60000	.62705	.09186	.02027
.70000	.77590	.11828	.02182
.80000	.89008	.11540	.02325
.90000	.97682	.12761	.02791
1.00000	1.16410	.22435	.02578

PO = 50

σ^2	$E\hat{\sigma}_n^2$	$E\lvert\hat{\sigma}_n^2-\sigma^2\rvert$	R_n
.10000	.10037	.01756	.04707
.20000	.19412	.03656	.01285
.30000	.29916	.04876	.01697
.40000	.46928	.08660	.02016
.50000	.52299	.06854	.02396
.60000	.66153	.07777	.02011
.70000	.81213	.16761	.01933
.80000	.80546	.12806	.02075
.90000	.85664	.15755	.02802
1.00000	1.11312	.15802	.02269

PO = 5000

σ^2	$E\hat{\sigma}_n^2$	$E\lvert\hat{\sigma}_n^2-\sigma^2\rvert$	R_n
.10000	.09567	.01631	.03623
.20000	.20296	.05674	.01928
.30000	.35087	.04815	.01905
.40000	.38703	.09775	.01792
.50000	.57548	.09917	.01502
.60000	.67951	.08766	.02315
.70000	.88786	.18198	.02394
.80000	.86060	.16171	.02136
.90000	.94229	.10485	.02831
1.00000	1.15474	.17834	.02648

The estimator $\hat{\sigma}_n^2$ is better than $\tilde{\sigma}_n^2$ as we can see for $\sigma^2 = 0.1$. For $\hat{\sigma}_n^2$ the case $p_0 = 50$ is the most satisfying in theory and in practice. Thus we shall estimate σ^2 with $\hat{\sigma}_n^2$ and $p = p_0 = 50$.

3.2. Dispersion of $\hat{\sigma}_n^2$

We will use the same function f, $\sigma^2 = 1$, $n = 1000$ and $p = 50$. We can see that the bias calculated is greater for $p_0 = 1000$ in Table 4. Table 5 shows that it is also more sparse.

$\hat{\sigma}_n^2$	
$p_0 = 100$	$p_0 = 1000$
1.1762	0.8055
0.9091	0.9880
1.1096	1.1226
0.8484	1.0703
0.8355	1.0251
0.7761	0.9821
0.9587	0.9285
0.9462	0.8331
0.9615	0.7578
0.8998	0.8253
1.2263	1.0955
1.1765	1.0077
1.0872	1.0150
1.0027	1.0953
0.9105	1.3566
0.9357	1.2831
1.0792	0.9821
0.9967	0.8709
1.0486	1.1834
0.9264	1.1023
$E\hat{\sigma}_n^2 = 0.990$	$E\hat{\sigma}_n^2 = 1.0165$

Table 4

Precision		1 %	5 %	10 %	15 %	20 %
frequences	PO = 100	0.10	0.25	0.60	0.70	0.90
	PO = 1000	0.05	0.30	0.50	0.60	0.85

Table 5

3.3. Study of $\hat{\sigma}_n^2$ for different regression functions

We display graphs showing the behaviour of $E\hat{\sigma}_n^2$ for different functions :

$f_1(x) = x \,/\, (1 + |x|^3)$

$f_2(x) = x$ here the estimates $E\hat{\sigma}_n^2$ are "infinite" because of a central limit effect.

$$f_3(x) = \begin{cases} e^{-2x^2} & ; x \geq 0 \\ (1+4x) \,/\, (1+x^2) & ; x < 0 \end{cases}$$

$f_4(x) = \sin 10x$

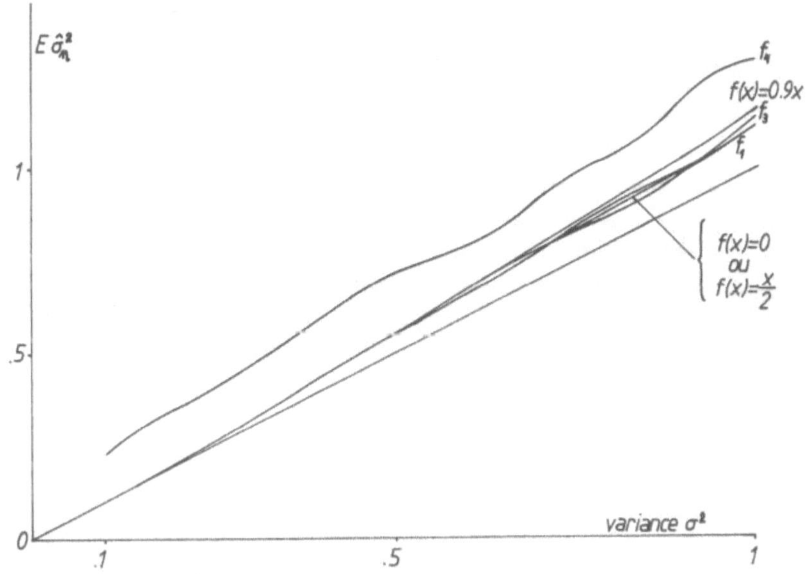

Figure 5

We see that the curves associated with f_1, f_3 and $f(x) = 0.05x$ or $0.9x$ are quite indistinguishable. The case $f(x) = 0$ is not the best (see $\sigma^2 = 0.6$). In the case of f_4 the variance is overestimated because this function varies a great deal. The estimation of the variance is good; in all the cases it is overestimated so that prediction could be carried out with this method.

4. VALUE OF THE ESTIMATE OF f AS A FUNCTION OF σ^2

4.1. Let us recall that $R_n \leq \frac{\alpha}{\sigma^4} + \beta\sigma^2$ for some scalars α and β[2]. Let us study R_n in the case of f_1, $n = 1000$ (Table 6). For large values of σ^2, R_n seems to be linear; this confirms the inequality given above ... but it seems to converge for very small values of σ^2. This can be explained as follows : the computer makes a truncation error η_n so that the process is really defined by $X_{n+1} = f(X_n) + \varepsilon_n + \eta_n$, where $\varepsilon_n \sim N(0,\sigma^2)$ and η_n is the internal white noise of the computer. Table 6 then allows us to evaluate $\text{Var}\ \eta_n \simeq 10^{-8}$ for the computer IRIS80 used for this work.

σ^2	10^8	10^6	10^4	10^2	10^0	10^{-2}	10^{-4}	10^{-6}	10^{-8}	10^{-10}
R_n	0.45×10^8	0.68×10^6	9622	13.75	0.01948	0.01753	0.2657	0.3314	0.3302	0.3304

Table 6

Figure 6 shows the estimates of the regression function in the cases f_2(a) and f_4(b) if $n = 1000$ and $\sigma^2 = 0.1$. For the identity function f_2, the mean $E\hat{f}_n$ satisfactorily estimates f but the realization $\hat{f}_n$ is far from f; this explains the divergence of the estimator of the variance. This is natural because, here, (X_n) is a recurrent random walk. For the sinusoidal function f_4, observe that the function is underestimated; this re-establishes an equilibrium because we have seen that the variance is overestimated; we may then think that, even in that bad case prediction would not be too bad.

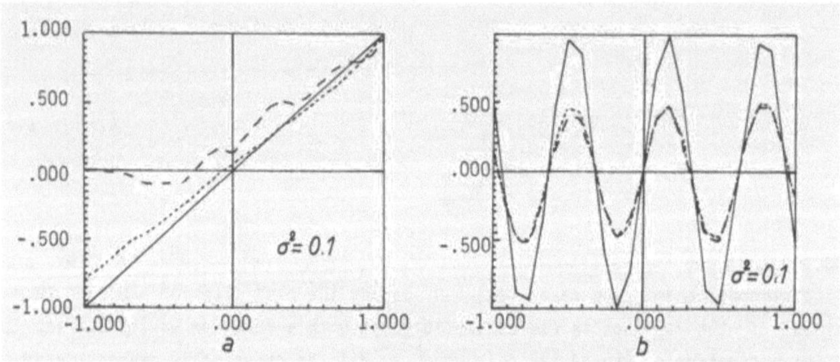

Figure 6

4.2. Study of the function $\sigma^2 \to R_n$

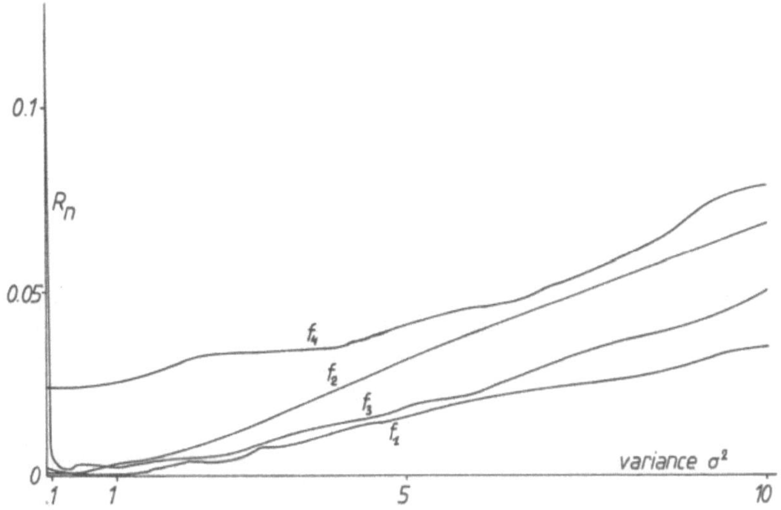

Figure 7

Except for the vicinity of the origin, which has already been interpreted (4.1) we can then see that, according to the theory, the function R_n is reasonably bounded by an expression such as : $\alpha\sigma^2 + \frac{\beta}{\sigma^4}$ [2].

4.3. A graphical representation

In practice, there are some autoregressive phenomena for which the variance can be increased, in particular in signal theory for telecommunications. It is therefore interesting to study the evolution of f_n with σ^2. For this, we use the function $f(x) = 1 / (1 + x^2)$ which satisfies the theoretical hypotheses. For x in the interval $[-1,1]$, $f(x) \geqslant 0.5$, so that X_n will very often take positive values and f will be badly estimated for small values of σ^2. Then, in order to obtain an optimal estimate of f we will have to increase σ^2. Here $n = 1000$.

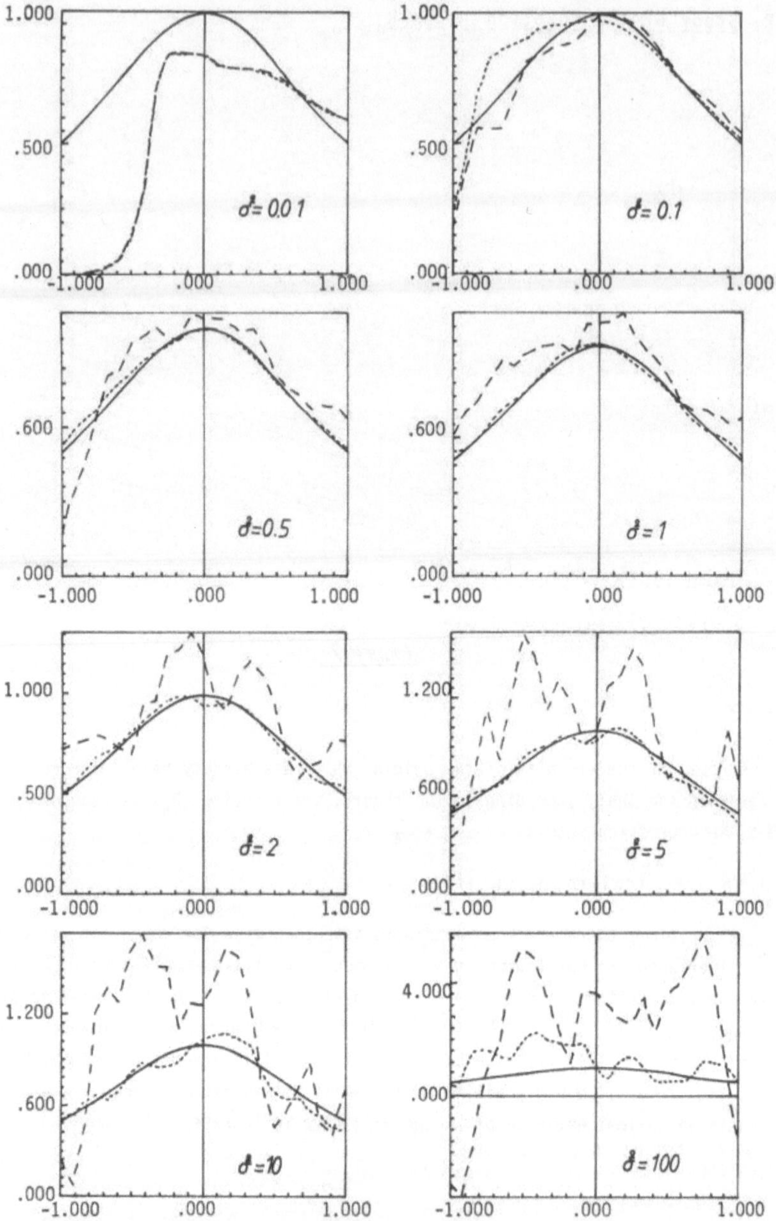

Figure 8

5. STUDY OF THE INVARIANT LAW WHEN σ^2 TENDS TO ZERO

5.1. *THEOREM D* [4]

Under the hypothesis of Theorem B, the process (X_n) *admits an invariant measure* Π_σ. *The limit points, for the vague convergence topology of* Π_σ *when* σ^2 *tends to zero admit a support included in the set :* $\bigcap_{n=0}^{+\infty} f^n(\mathbb{R})$. *Moreover, if* f *is continuous they are invariant with respect to the function* f.

We shall study this behaviour by simulations. For this we use two categories of graphical representations. We shall display on the left, representations of the estimates of f as made before and, on the right, we represent $\hat{\Pi}_n$.

- The continuous line represents the function $x \to E^{X_0=0}(\hat{\Pi}_n(x))$.
- The dotted line and the interrupted line represent two realizations of $x \to \hat{\Pi}_n(x)$.

In Figure 9,

$$f(x) = \begin{cases} 0, & x \geq 0 \\ -1, & x < 0 \end{cases} \qquad \sigma^2 = 10^{-8} \; ; \; n = 1000.$$

The limit distribution is here the Dirac measure at (-1) as is shown by Theorem D.

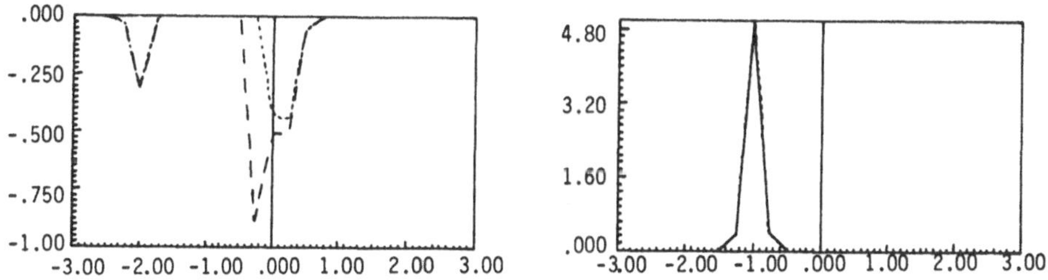

Figure 9

5.2. Iteratively convergent functions

For such functions the limit distribution of Π_σ is a Dirac measure at the fixed point of f as it is shown by Theorem D. In [6], F. Guénard shows a non asymptotic analogous result for a compact supported white noise. Figure 10 shows the importance of the continuity of f, here

$$f(x) = \begin{cases} 0, & x < 0 \\ -1, & x \geq 0 \end{cases}$$

For this case, we show [4] that the limit distribution is $\frac{1}{3}(2\delta_0 + \delta_{-1})$, as is confirmed by simulation. This limit is not invariant with respect to f.

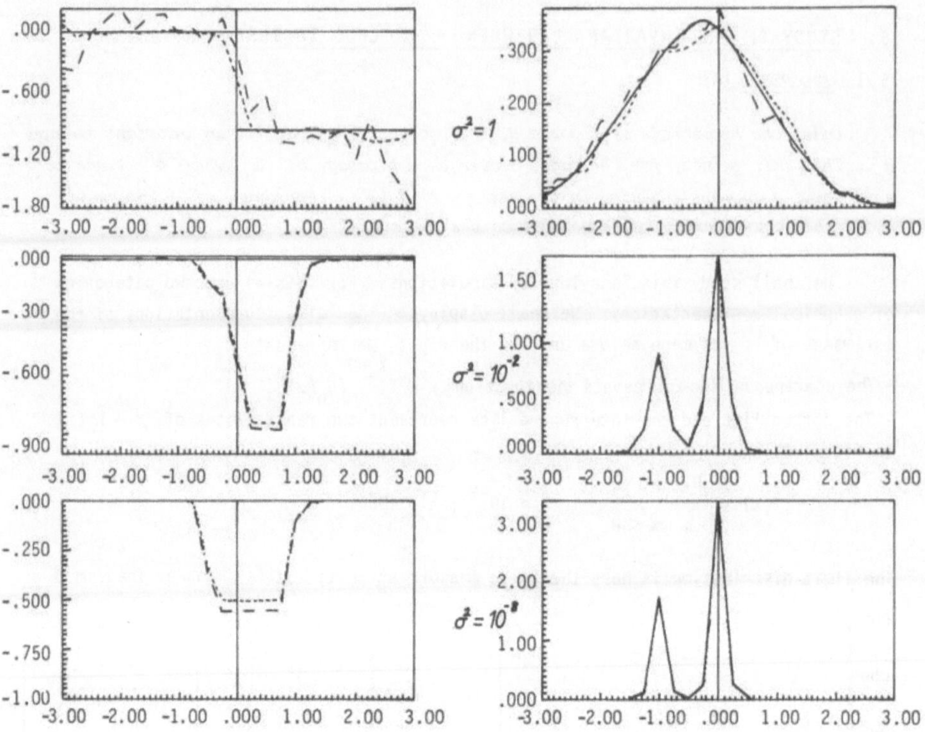

Figure 10

Now the case of Figure 11 shows a good example, here $f(x) = \frac{\alpha}{\sin\alpha}\sin x$ where $\alpha \in [\Pi/2,\Pi]$ and $\alpha + \operatorname{tg}\alpha = 0$. This case is interesting because f has three fixed points; 0 is repulsive, α and $-\alpha$ are attractive and $f'(\alpha) = -1$ so that the speed of convergence is not geometric [5]. As shown by Theorem D the limit of Π_σ is a linear combination of the Dirac-measures at 0, α and $-\alpha$. See [6] for further interpretations of this case.

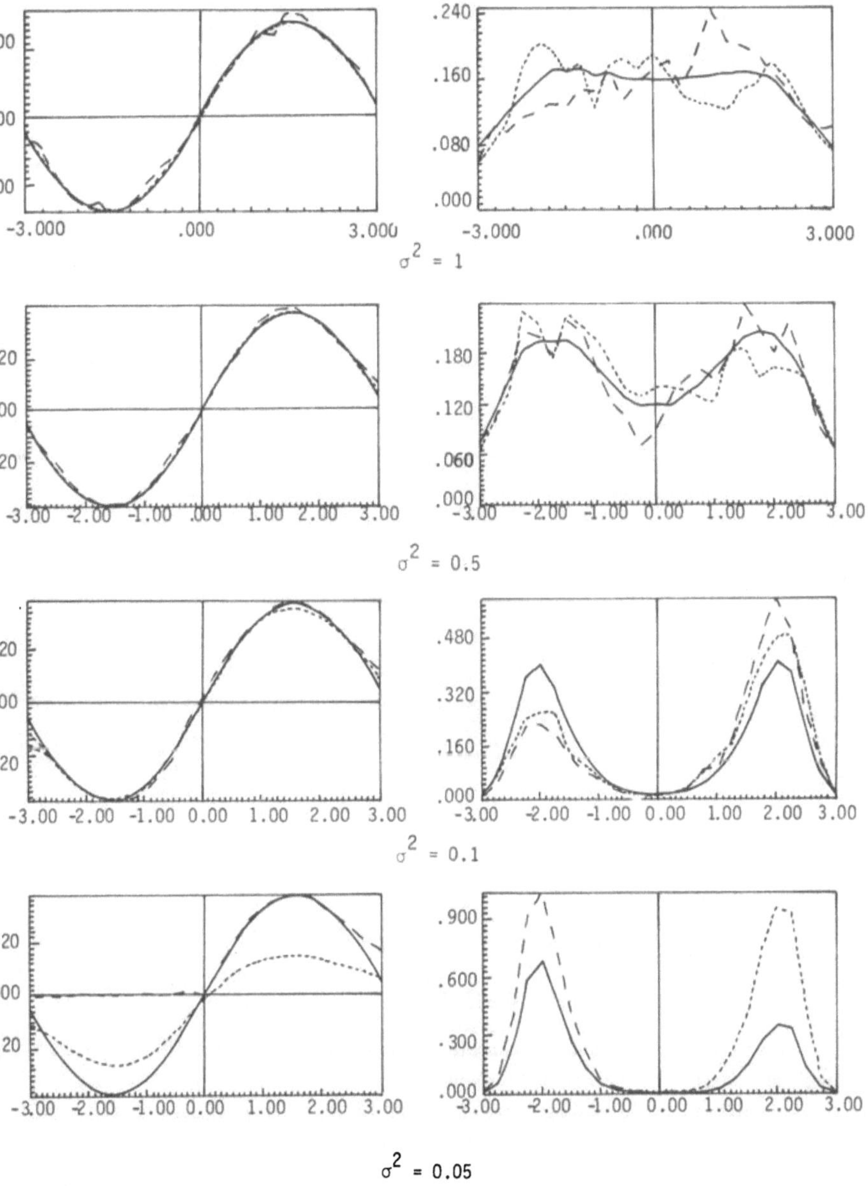

1.600
.000
-1.600
-3.000
.000
3.000
.240
.160
.080
.000
-3.000
.000
3.000
σ2 = 1
1.20
.000
-1.20
-3.00 -2.00 -1.00 .000 1.00 2.00 3.00
.180
.120
.060
.000
-3.00 -2.00 -1.00 0.00 1.00 2.00 3.00
σ2 = 0.5
1.20
.000
-1.20
-3.00 -2.00 -1.00 0.00 1.00 2.00 3.00
.480
.320
.160
.000
-3.00 -2.00 -1.00 0.00 1.00 2.00 3.00
σ2 = 0.1
1.20
.000
-1.20
-3.00 -2.00 -1.00 0.00 1.00 2.00 3.00
.900
.600
.300
.000
-3.00 -2.00 -1.00 0.00 1.00 2.00 3.00
σ2 = 0.05

Figure 11

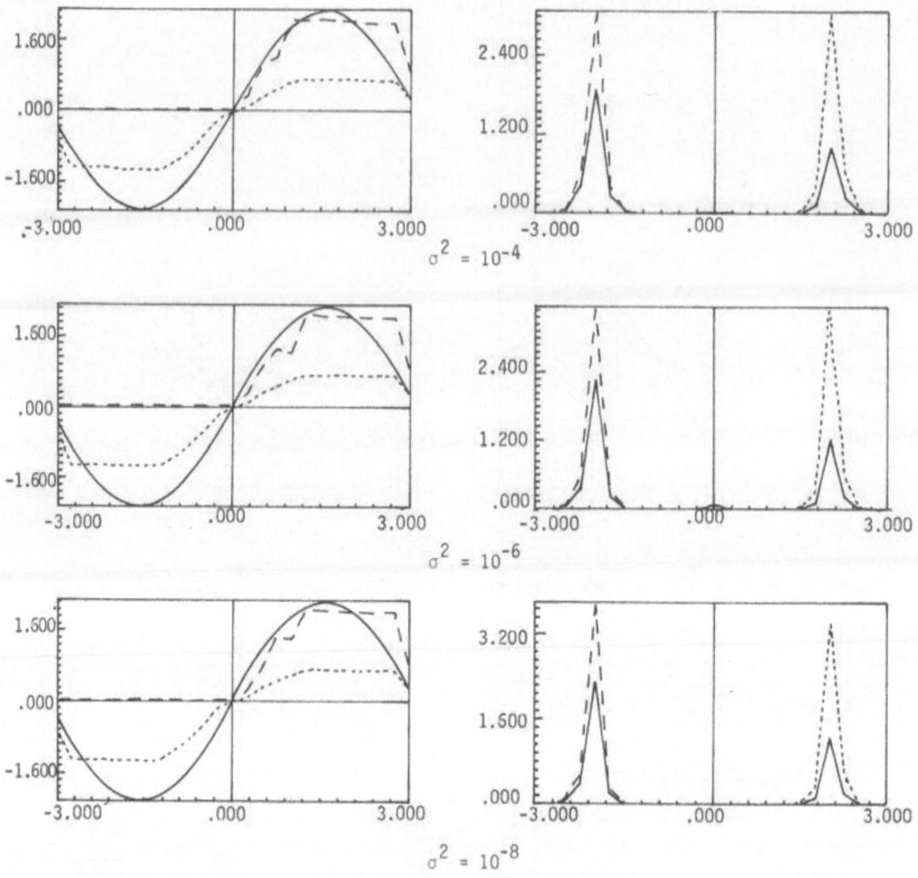

Figure 11 (fin)

5.3. A function having an absolutely continuous invariant function

D. Ruelle [7] gave such an example : $f(x) = 4x(1-x)$, $x \in [0,1]$. Here an invariant measure with respect to f on $[0,1]$ has the density $\frac{1}{\Pi}\frac{1}{\sqrt{x(1-x)}}$. So if we define f by 0 on the complement of $[0,1]$ a linear combination of this measure and of the Dirac measure at 0 is invariant with respect to f. Figure 12 shows this phenomenon.

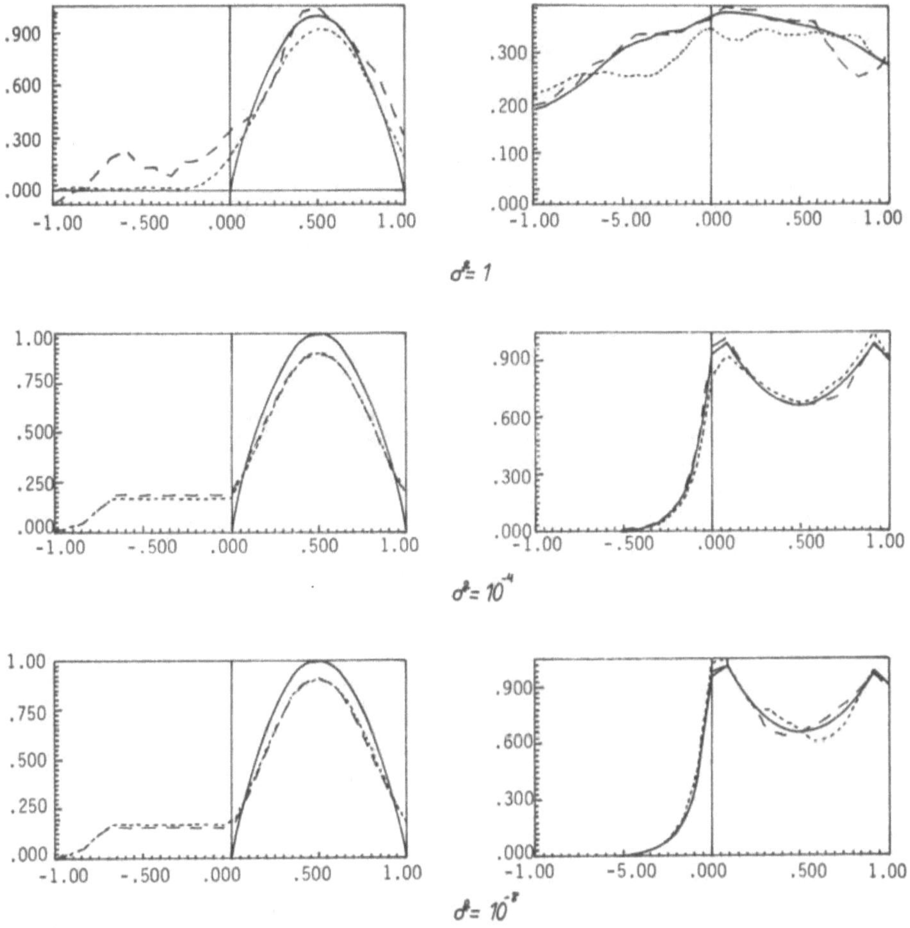
$\sigma^2 = 1$
$\sigma^2 = 10^{-4}$
$\sigma^2 = 10^{-8}$

Figure 12

REFERENCES

[1] Doukhan, P. and M. Ghindès (1980), *C.R.A.S. Série A*, t. 290, pp. 921-923.

[2] Doukhan, P. and M. Ghindès (1980), *C.R.A.S. Série A*, t. 291, pp. 61-64.

[3] Doukhan, P. and M. Ghindès (1980), "Estimation de la transition de probabilité d'une chaîne de Markov Doeblin récurrente; étude du cas particulier du processus autorégressif d'ordre 1", *Prépublication d'Orsay 80T55*, to appear in *Stochastic Processes and their Applications* (1982).

[4] Doukhan, P. (1980), Thèse de troisième cycle, Université d'Orsay n° 2859.

[5] Guénard, F. (1981), *C.R.A.S. Série A*, t. 292, pp. 55-58.

[6] Guénard, F. (to appear 1982), "Itérations stochastiques et déterministes sur les intervalles", *Proc. Conf. of non Linear Analysis Saint Jones*, Juin 1981, Academic Press.

[7] Ruelle D. (1977), "Applications conservant une mesure absolument continue par rapport à dx sur [0,1], *Commentat. Phys.-Math.* 55, pp. 47-51.

NON PARAMETRIC PREDICTION IN STATIONARY PROCESSES

by

Denis Bosq

University of Lille I - France

Abstract

We present a class of non parametric predictions and study their asymptotic properties. The general framework of the study allows us to predict the distribution of the nonobserved variable. We give also the results of some simulations carried out in an elementary case.

Key-words : Nonparametric statistics, Prediction theory, Stationary Processes, Asymptotic theory

AMS/MOS : Primary 62G99, Secondary 60G10, 62M20

NON-PARAMETRIC PREDICTION IN STATIONARY PROCESSES

by Denis BOSQ

(University of Lille I - France).

I - INTRODUCTION.-

The general prediction problem can be stated as follows "Predict Y, a nonobserved random variable, from X, an observed random variable".

The solutions of classical prediction problems are well known (conditional expectation, linear predictor) but it is important to distinguish between the statistical and the probabilistic solutions : they correspond to two errors of prediction ; indeed, if Y is real, if $R(X)$ is the best predictor and $\hat{R}(X)$ its statistical approximation, the prediction error is given by the formula

$$E\left[Y - \hat{R}(X)\right]^2 = E\left[Y - R(X)\right]^2 + E\left[R(X) - \hat{R}(X)\right]^2$$

If the model chosen by the statistician is correct the problem consists only in obtaining a small value for $E\left[R(X) - \hat{R}(X)\right]^2$; if not, the statistician works on a wrong value of R and the accuracy of $\hat{R}$ is illusory if this predictor does not possess robustness properties.

This kind of drawback is perfectly illustrated by the BOX and JENKINS method : the choice of an ARMA model is rather arbitrary and the maximum likelihood estimates of the parameters are not robust so that the predictor is sometimes disappointing.

Therefore the use of nonparametric methods offers advantages : a nonparametric predictor is simple to calculate, robust and sufficiently precise.

In this paper we present a class of nonparametric predictors and study their asymptotic properties. The general framework of the study allows us to predict the distribution of Y. In the last paragraph we give the results of some simulations carried out in an elementary case.

II - A CLASS OF NONPARAMETRIC PREDICTORS.-

1 - Let $(\xi_t, t \in \mathbb{Z})$ be a strictly stationary Markovian process taking values in $]0,1[$.

Set $B_j = \left]\frac{j}{p}, \frac{j+1}{p}\right]$; $j = 0,\dots,p-1$ and write $B_j(\xi_t)$ for the interval B_j which contains ξ_t ; $t = 0,\dots,n$. When $X = (\xi_0,\dots,\xi_n)$ and $Y = \xi_{n+k}$, $k > 0$ one can construct a simple and natural nonparametric predictor : the predictogram.

Put $\hat{R}(X) = 0$ if $\sum_{t=0}^{n-k} 1_{B_j(\xi_n)}(\xi_t) = 0$

and

$$\hat{R}(X) = \left[\sum_{t=0}^{n-k} 1_{B_j(\xi_n)}(\xi_t)\right]^{-1} \sum_{t=0}^{n-k} 1_{B_j(\xi_n)}(\xi_{t+k})$$

otherwise.

When the distribution μ of ξ_0 is known and such that

$$p_j = \mu(B_j) > 0 \qquad j = 0,\dots,p-1$$

we can define the modified predictogram by the formula

$$\hat{R}^*(X) = \frac{1}{n-k+1} \sum_{t=0}^{n-k} K(\xi_t, \xi_n)\, \xi_{t+k} \tag{1}$$

where

$$K(x,y) = \sum_{j=1}^{p} p_j^{-1}\, 1_{B_j}(x)\, 1_{B_j}(y) \quad ; \quad x, y \in]0,1[$$

Formula (1) suggests the general form of the predictors we want to study.

2 - More generally, let $(\xi_t, t \in \mathbb{Z})$ be a (strictly) stationary process, defined on a probability space $(\Omega,\mathcal{A},P)$, and taking values in a measurable space $(E,\mathcal{B})$. μ, the distribution of ξ_0, is supposed to be known, but not the conditional distributions between the ξ_t's.

Let g be a transformation from E into H, a separable Hilbert space. We want to predic $g(\xi_{n+k})$ from the observations $\xi_0,\dots,\xi_n$. More precisely, if $g \in L^2_H(\mu)$, we want to approach $E(g(\xi_{n+k})|\xi_n)$ in the $L^2_H(P)$ norm by a function of $\xi_0,\dots,\xi_n$. Of course, $E(g(\xi_{n+k})|\xi_n)$ is the best predictor of $g(\xi_{n+k})$ only when (ξ_t) is Markovian, but, from a mathematical point of view, we do not need any hypothesis of Markovian type. It should be noticed that the prediction of $g(\xi_{n+k})$ can be interpreted as an estimation of the random variable $E[g(\xi_{n+k})|\xi_n]$.

Then, let us consider a sequence $(K_n, n \geq k)$ of real functions defined on $E \times E$, $\mathcal{B}^2$-measurable, symetric and bounded. We set, for $n \geq k$,

$$\hat{R}_n = \hat{R}_n(\xi_o,\ldots,\xi_n) = \frac{1}{n-k+1} \sum_{t=0}^{n-k} K_n(\xi_t, \xi_n)\, g(\xi_{t+k}) .$$

In the sequel we shall investigate the asymptotic behaviour of $\hat{R}_n$ when n tends to infinity.

3 - In order to "measure" the ergodicity of the process (ξ_t), we shall use the mixing coefficient introduced by J. GASTWIRTH and H. RUBIN in [4] : let W and Z be two random variables taking values in F and G respectively and such as a conditional distribution P_Z^W of Z fiven W exists. We put

$$\nu(y) = P_Z^{W=y} - P_Z , \quad y \in F$$

and we denote by $\Delta(y)$ the total variation of $\nu(y)$.

Then the following result is a straightforward generalisation of the GASTWIRTH and RUBIN lemma :

Lemma.- If H is a separable Hilbert space and if $\varphi(W) \in L_H^q(\Omega,A,P)$ and $\Psi(Z) \in L_H^r(\Omega,A,P)$ where q and $r \in [1, +\infty]$, then

$$|E <\varphi(W) - E\varphi(W), \Psi(Z) - E\Psi(Z)>_H| \leq 2^{1/q} . ||\varphi(W)||_{L_H^q} . ||\Psi(Z)||_{L_H^r} . D$$

where $D = ||\Delta^{1/r + 1/s}||_{L^s(P_W)}$ with $\frac{1}{q} + \frac{1}{r} + \frac{1}{s} = 1$.

We apply this lemma to the random variables $W = (\ldots,\xi_{t-1},\xi_t)$ and $Z = (\xi_{t+p}, \xi_{t+p+1}, \ldots)$ for which we suppose existence of P_Z^W. The corresponding D will be denoted by $D_p(r,q)$.

Now we can state (E), the ergodicity hypothesis for (ξ_t) :

(E) $\quad \sum_{p=0}^{\infty} p\, D_p(q,q) < +\infty$ for some $q \geq 2$.

This mixing condition is satisfied in many familiar cases, in particular by autoregressive processes.

4 - In this paragraph, we make assumptions about the sequence (K_n). First, K_n is the kernel of a symetric Hilbert Schmidt operator on $L^2(\mu)$, consequently there exists an orthonormal sequence $(\varphi_{jn}, j \in J_n)$ and a sequence of real numbers $(\lambda_{jn}, j \in J_n)$ such that

$$K_n(x,y) = \sum_{j\in J_n} \lambda_{jn} \varphi_{jn}(x) \varphi_{jn}(y) \quad ; \quad x , \; y \in E$$

where $J_n = \{1,\dots,d_n\}$ or $J_n = \mathbb{N}$. The series converges in $L^2(\mu)$-norm.

We assign to (K_n) the additional condition :

(N) $$\sum_{j\in J_n} |\lambda_{jn}| \sup_{x\in E} |\varphi_{jn}(x)|^2 < +\infty , \quad n \geq k .$$

Moreover, let $\mathcal{R}$ the family of possible regressions of $g(\xi_k)$ on ξ_o. We suppose that $\mathcal{R} \subset L^2_H(\mu)$ and we put

$$(K_n R)(x) = \int K_n(x,t) R(t) \, d\mu(t) \quad ; \quad x \in E \quad , \quad R \in \mathcal{R}$$

where the integral is taken in the Bochner sense.

Now we can introduce an hypothesis which asserts that (K_n) tends to the identity :

(I) $$\lim_{n\to+\infty} ||K_n R - R||^2_{L^2_H(\mu)} = 0 \quad , \quad R \in \mathcal{R} .$$

5 - The first proposition provides a bound for the statistical prediction error $\Delta_n = E\,||\hat{R}_n - R(\xi_n)||^2_H$:

<u>Proposition</u> 1.- If (N) is satisfied and if $g \in L^q_H(\mu)$, then

$$\Delta_n \leq 2\,||K_n R - R||^2_{L^2_H(\mu)} + \frac{\alpha_n}{n-k+1} \sum_j \lambda^2_{jn} M^4_{jn} + \frac{\beta_n}{n-k+1} \sum_{j\neq j'} |\lambda_{jn} \lambda_{j'n}| M^2_{jn} M^2_{j'n} \qquad (2)$$

where $M_{jn} = \sup_{x\in E} |\varphi_{jn}(x)|$

$$\alpha_n = 2\,||g||^2_{L^q_H(\mu)} \left[1 + 2(k+1) + 2^{1+q^{-1}} \sum_{p=0}^{n-2k} D_p(q,q)\right]$$

and

$$\beta_n = 16\,||g||^2_{L^q_H(\mu)} \left[2 \sum_{p=0}^{[n/2]-k} (p+2)\, D_p(q,q) + \sum_{p=0}^{n-k} (p+k+\frac{1}{2})\, D_p(\frac{q}{2}, +\infty)\right]$$

The proof, which is rather long and technical, uses essentially the inequality in the lemma. The reader is refered to [2] for a complete proof.

6 - The following convergence theorem is an easy consequence of proposition 1.

Proposition 2.- Under the conditions (E) ; (N) ; (I) ; $g \in L^q_H(\mu)$; $\lim_{n\to\infty} \frac{1}{n} \sum_j |\lambda_{jn}|^\beta M_{jn}^{2\beta} = 0$; $\beta = 1,2$ we have $\lim_{n\to\infty} \Delta_n = 0$.

7 - A special case : If $K_n = \sum_{j=1}^{d_n} \varphi_j \otimes \varphi_j$, $n \geq 1$ where $(\varphi_j, j \geq 1)$ is a uniformly bounded orthonormal basis of $L^2(\mu)$ and $(d_n) \to +\infty$, we have the following result :

Proposition 3.- Under the conditions (E) and $g \in L^q_H(\mu)$, $\lim_{n\to+\infty} \Delta_n = 0$ if and only if $\lim_{n\to+\infty} \frac{d_n}{n} = 0$.

Furthermore

$$\Delta_n = O\left(\sum_{j>d_n} \left\| \int \varphi_j \; R \; d\mu \right\|_H^2\right) + O\left(\frac{d_n}{n}\right) \tag{3}$$

The "if" part is a special case of proposition 2. To prove the "only if" part it suffices to consider the i.i.d. case. Finally (3) is a consequence of (2).

III - APPLICATIONS.-

1 - The preceding results apply directly to the prediction of a $\mathbb{R}^d$ - valued random variable. Furthermore the predictor is efficient even when the model is parametric : for instance, if μ is a probability on $]0,1[$ and R a polynomial of degree $\leq D$ one can build a predictor of ξ_{n+k} based upon the reproducing kernel K of the subspace of $L^2(\mu)$ generated by the polynomials of degree $\leq D$. Then we have

$$\hat{R}_n = \frac{1}{n-k+1} \sum_{t=0}^{n-k} K(\xi_t, \xi_n) \cdot \xi_{t+k} \tag{4}$$

and the statistical prediction error is $O(\frac{1}{n})$.

2 - It is often more interesting to use a predictor which estimates the entire conditional distribution of ξ_{n+k} rather than a predictor which estimates only its conditional expectation. We give now three examples of predictors of this type :

a) "Estimation" of the conditional distribution function.

If $(E,\mathcal{B}) = ([0,1], \mathcal{B}_{[0,1]})$ and if $g(x) = 1_{[x,1]}(.)$, $x \in [0,1]$ then $g \in L^2_{L^2(\lambda)}(\mu)$ where λ is the Lebesgue measure. It is easy to see that

$$E\left[g(\xi_k)(.)|\xi_o\right] = P\left[\xi_k \leq . |\xi_o\right]$$

If $\mu = \lambda$ we can build a predictor of $g(\xi_{n+k})$ based upon trigonometric functions ;

we put

$$K_n(x,y) = 1 + 2 \sum_{j=1}^{h_n} (\cos 2\pi\ jx \cos 2\pi\ jy + \sin 2\pi\ jx \sin 2\pi\ jy)$$

hence

$$K_n(x,y) = \frac{\sin 2\pi \frac{2h_n+1}{2}(y-x)}{\sin 2\pi \frac{y-x}{2}} \quad \lambda^2 \text{ a.e.}$$

then, if $P(\xi_n = \xi_t) = 0$; $t = 0,\dots,n-k$,

$$\hat{R}_n(u) = \frac{1}{n-k+1} \sum_{t=0}^{n-k} \frac{\sin 2\pi \frac{2h_n+1}{2}(\xi_n-\xi_t)}{\sin 2\pi \frac{\xi_n-\xi_t}{2}} 1_{[\xi_{t+k},1]}(u) , \qquad u \in [0, 1].$$

This formula is valid for every u, almost surely.

By proposition 3 one can establish that, under condition (E), $\lim_{n\to\infty} \Delta_n = 0$ if and only if $\lim_{n\to\infty} \frac{h_n}{n} = 0$. In addition, we have

$$\Delta_n = O\Big(\sum_{j>h_n} \int_0^1 \Big[\int_0^1 \cos 2\pi\ jx\ F(y|x)\ dx\Big]^2 dy + \sum_{j>h_n} \int_0^1 \Big[\int_0^1 \sin 2\pi\ jx\ F(y|x)\ dx\Big]^2 dy + \frac{h_n}{n}\Big)$$

where $F(.|.)$ is the conditional distribution function of ξ_k relative to ξ_o .

b) <u>"Estimation" of the conditional distribution.</u>

In this paragraph we assume that $(E,\mathcal{B})$ is a polish space equipped with its Borel σ-field. Let $M(\mathcal{B})$ be the space of bounded signed measures defined on $(E,\mathcal{B})$ and $P(\mathcal{B})$ the space of probabilities on $(E,\mathcal{B})$. In [5] C. GUILBART has shown that there exists a bounded $\mathcal{B} \otimes \mathcal{B}$ - measurable reproducing kernel , say N, such that

$$<\lambda_1,\lambda_2> = \int N(x,y)\ d\lambda_1(x)\ \lambda_2(y) \quad ; \quad \lambda_1 , \lambda_2 \in M(\mathcal{B})$$

is a scalar product on $M(\mathcal{B})$ which induces on $P(\mathcal{B})$ a topology compatible with the weak topology and which identifies $M(\mathcal{B})$ as a subspace of H_N , the reproducing kernel Hilbert space generated by N. Finally one can choose N such as H_N be separable.

Then, if we put $g(.) = \delta_{(.)}$, where $\delta_{(a)}$ denotes the Dirac measure on a, it can be seen that $g \in L^2_{H_N}(\mu)$ and that $E[g(\xi_k)|\xi_o]$ is the conditional distribution ν^{ξ_o} of ξ_k given ξ_o .

Let us consider the predictor

$$\hat{R}_n = \frac{1}{n-k+1} \sum_{t=0}^{n-k} K_n(\xi_t,\xi_n)\ \delta_{(\xi_{t+k})}$$

Under the conditions of proposition 2, we have

$$\lim_{n\to\infty} E||\hat{R}_n - \nu^{\xi_n}||^2_{H_N} = 0 \ .$$

For possible choices of N see [5]. If, for instance E is a compact interval of $\mathbb{R}$ we can take

$$N(x,y) = e^{xy} \quad ; \quad x\ ,\ y \in E\ .$$

c) "Estimation" of the conditional density.

The predictor defined by (4) is essentially of theoretical interest. In order to obtain a more practical predictor we consider $f(.|\xi_n)$, the conditional density of ξ_{n+k} given ξ_n. As the random variable $f(.|\xi_n)$ is not a conditional expectation, it is necessary to modify slightly the method of prediction. Thus we shall use a predictor of the form $(n-k+1)^{-1} \sum_{t=0}^{n-k} g_n(.,\ \xi_{t+k})\ K_n(\xi_t,\ \xi_n)_a$ where g_n is the kernel of some operator.

In order to define the predictor more explicitly we consider only the case where $(E,\mathcal{B},\mu) = ([0,1],\ \mathcal{B}_{[0,1]},\ \lambda)$ and we suppose that $f(.|.)$ is defined and continuous on $[0,1]^2$.

Now, let (η_n) be a sequence of strictly positive real numbers and (φ_j) a uniformly bounded orthonormal basis of $L^2(\lambda)$. We put

$$\hat{R}'_n = \frac{1}{\eta_n(n-k+1)} \sum_{t=0}^{n-k} 1_{].,.+\eta_n]}(\xi_{t+k}) \sum_{j=1}^{d_n} \varphi_j(\xi_t)\ \varphi_j(\xi_n) \ .$$

Then, it is easy to estallish the following result (see [2]).

Corollary.- If $\sum_p p\ D_p(+\infty,\ +\infty) < \infty$, $\lim_n h_n = 0$, $\lim_n d_n = +\infty$, $\lim n^{-1}\ h_n^{-1}\ d_n = 0$ then $\lim_n \Delta_n = 0$.

IV - THE GENERAL CASE.-

1 - In pratice, μ is only known approximatively or may even be completely unknown. In the first case the above method still works, in the second, modifications are necessary.

2 - The following proposition shows a certain robustness of $\hat{R}_n$ when μ varies :

Proposition 4.- Let μ_o be a probability on $(E,\mathcal{B})$ such that $\frac{d\mu}{d\mu_o}$ exists and satisfies $0 < a \leq \frac{d\mu}{d\mu_o} \leq b$ where a and b are constant.

Suppose that

1) On R, (K_n) converges strongly to the identity of $L^2_H(\mu_o)$.

2) (N) is satisfied (for the decomposition of (K_n) in $L^2(\mu)$).

3) $g \in L^q_H(\mu_o)$ and $\sum_p p\, D_p(q,q) < +\infty$.

4) $\lim_{n\to\infty} n^{-1} \sum_j |\lambda_{jn}|^\beta M^{2\beta}_{jn} = 0$; $\beta = 1 , 2$.

Then

$$\limsup \Delta_n \leq 2b \left|\left|\left(1 - \frac{d\mu}{d\mu_o}\right) R\right|\right|^2_{L^2_H(\mu_o)} \tag{5}$$

See [2] for a proof.

Consequently $\hat{R}_n$ is still a good predictor if the presumed distribution μ_o is close to the true distribution μ or at least if the right - hand side of (5) is small with respect to the probabilistic prediction error $||g||^2_{L^2_H(\mu)} - ||R||^2_{L^2_H(\mu)}$ (for (ξ_t) Markovian).

3 - In The general case one must use a predictor of the formm

$$\frac{\sum_{t=0}^{n-k} K_n(\xi_t,\xi_n)\, g(\xi_{t+k})}{\sum_{t=0}^{n-k} K_n(\xi_t,\xi_n)}$$

There are few mathematical results about this type of predictor but some simulations have shown that they compare well with parametric methods (see V).

For asymptotic results on the predictogram the reader is refered to [3].

V - SIMULATIONS.-

The simulations which appear below have been carried out by R. KALAIDJIAN ; the process (ξ_t) is a disturbed ARMA process.

The values of ξ_{n+k} are in continuous lines, the BOX and JENKINS predictor is chain dotted, the nonparametric predictors are in "broken" lines.

One can see that the performances of the nonparametric predictors are quite satisfactory even in the parametric situation (see [6] for the details).

BIBLIOGRAPHY.

[1] D. BOSQ - Une méthode non paramétrique de prédiction d'un processus stationnaire. Prédiction d'une mesure aléatoire. CRAS, A, t. 290, p. 711-713 (1980).

[2] D. BOSQ - Sur la prédiction non paramétrique de variables aléatoires et de mesures aléatoires. Publ. Interne - U.E.R. de Maths. Pures et Appl. Lille I - n° 164 (1979).

[3] G. COLLOMB - Convergence du prédictogramme en prédiction non paramétrique. (In this volume).

[4] J.L. GASTWIRTH - H. RUBIN - The asymptotic distribution theory of the empiric C.D.F. for mixing stochastic processes. Annals of Stat. Vol. 3 n° 4 p. 800-824 (1975).

[5] C. GUILBART - Etude des produits scalaires sur l'espace des mesures. Annales de l'I.H.P. (1979).

[6] R. KALAIDJIAN - Ebauche d'une étude comparée sur les efficacités de certaines méthodes de prédiction pour un ARMA perturbé "Statistique et Analyse des données"). (1982).

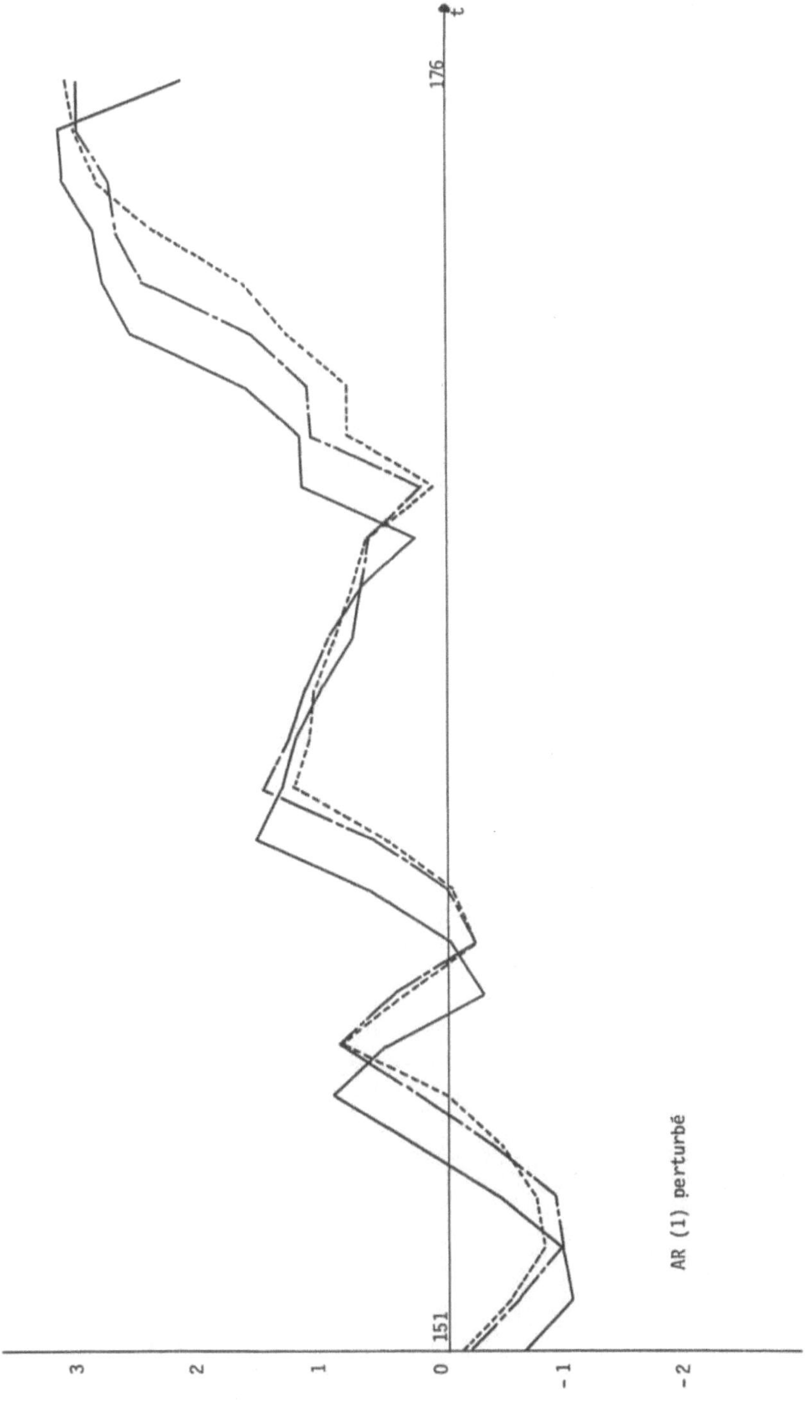
t
176
151
AR (1) perturbé
3
2
1
0
-1
-2

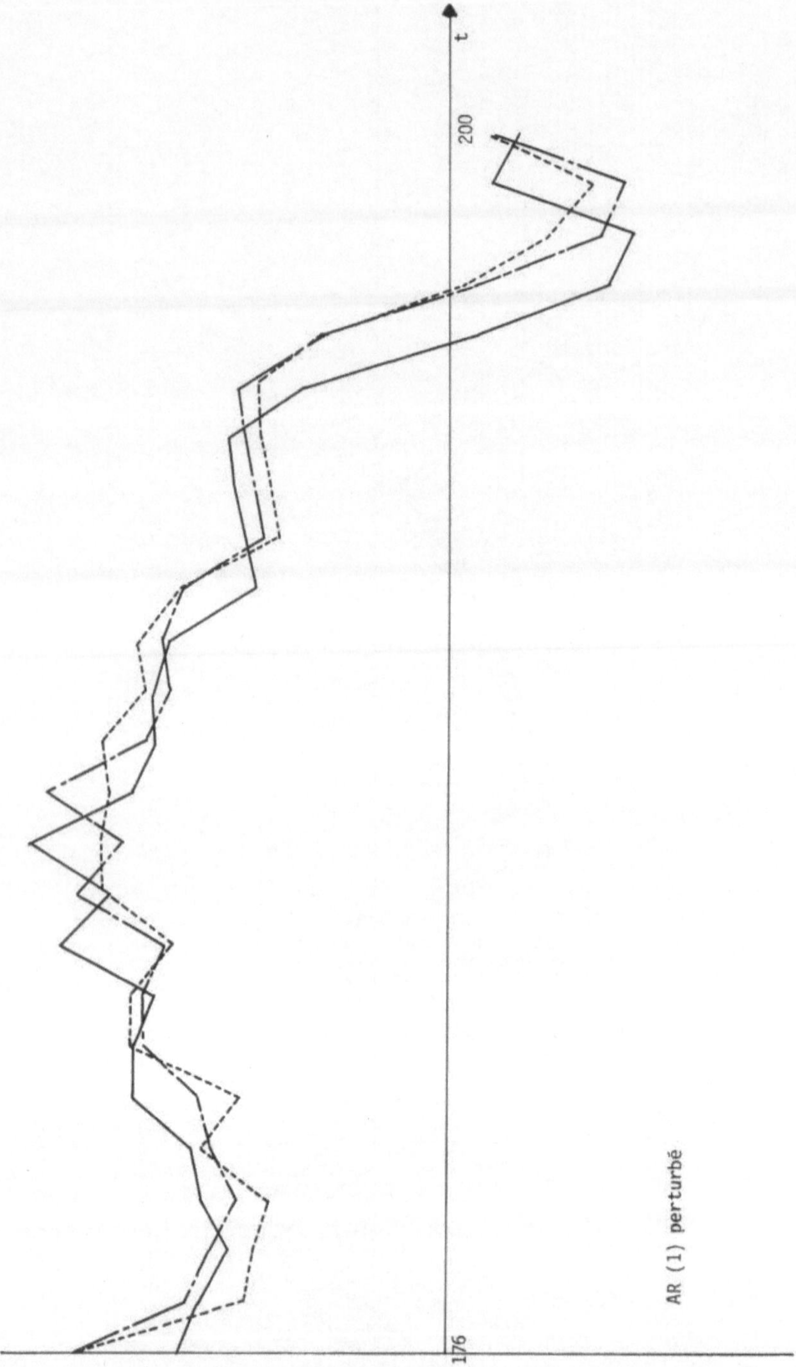
t
200
176
AR (1) perturbé

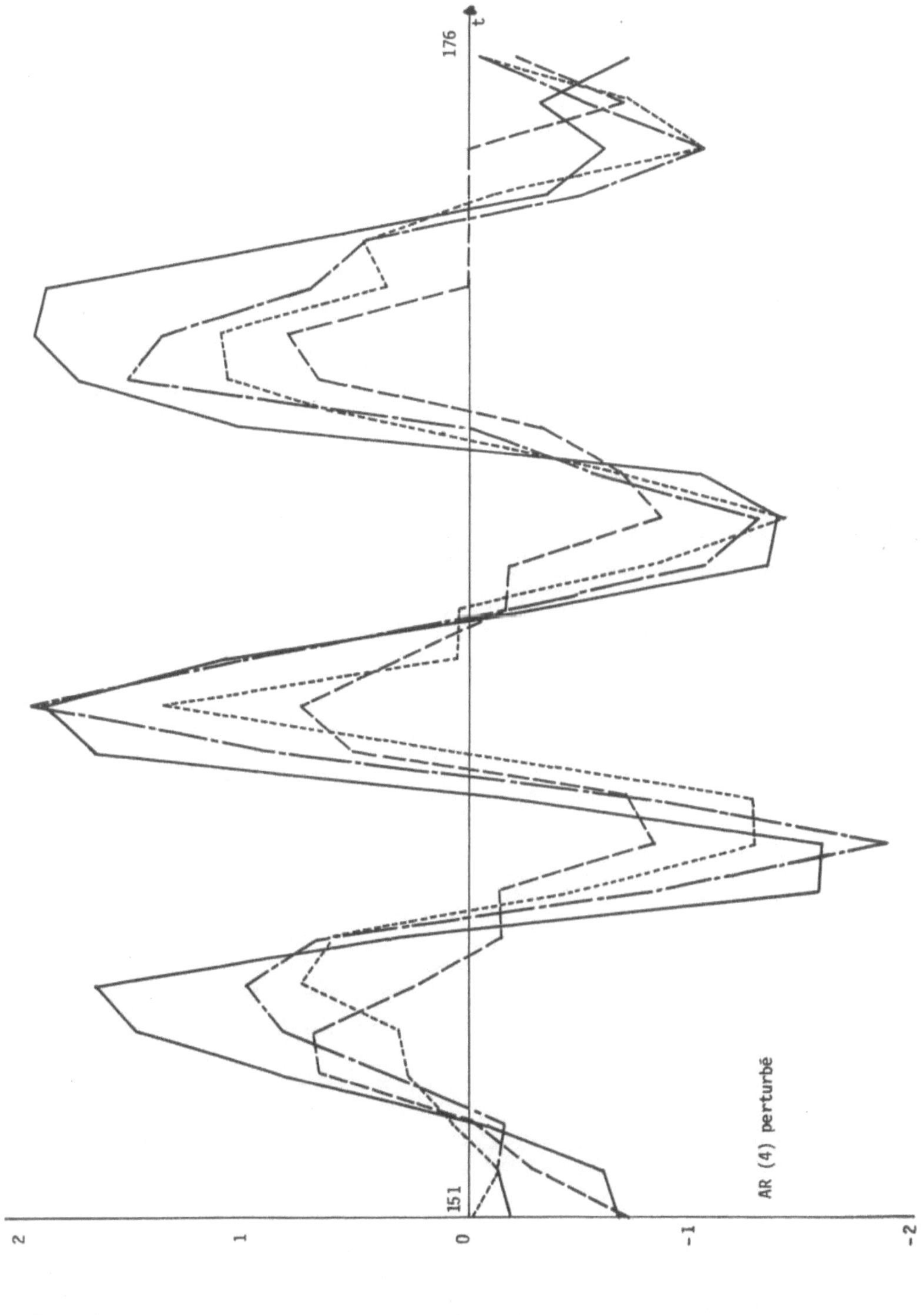

2
1
0
-1
-2
151
176
t
AR (4) perturbé

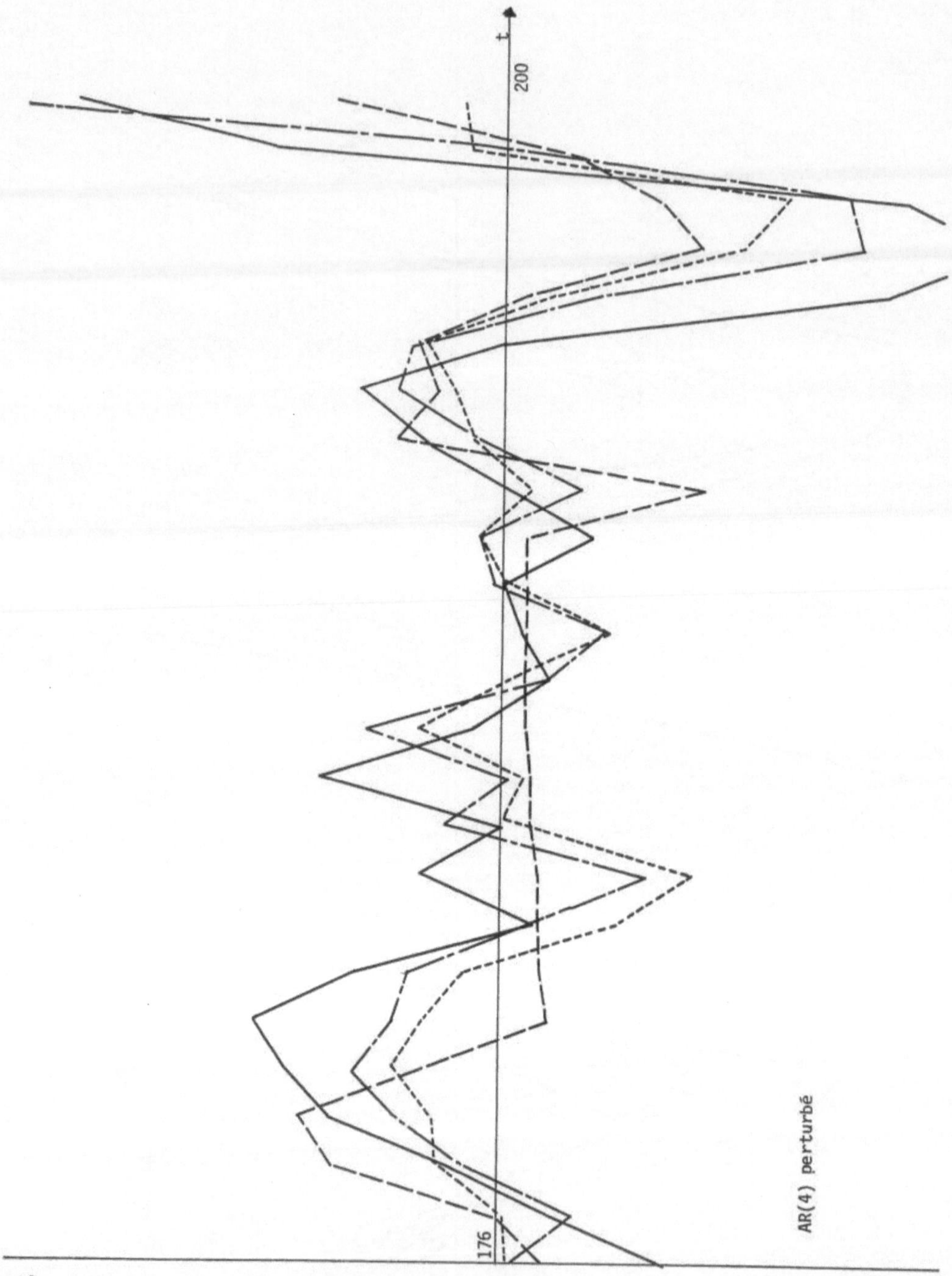

2
1
0
-1
176
200
t
AR(4) perturbé

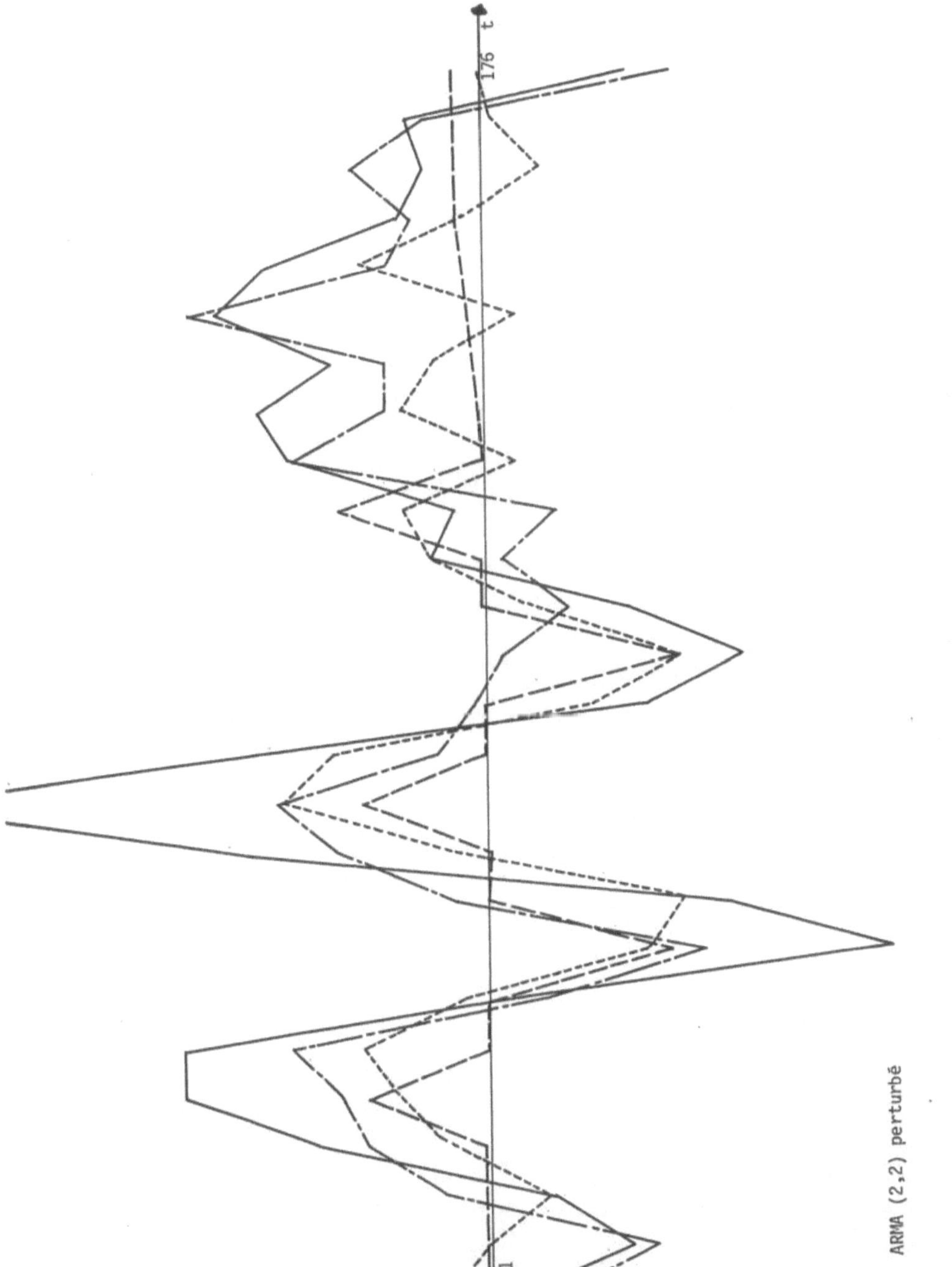

t
176
151
1
0
-1
ARMA (2,2) perturbé

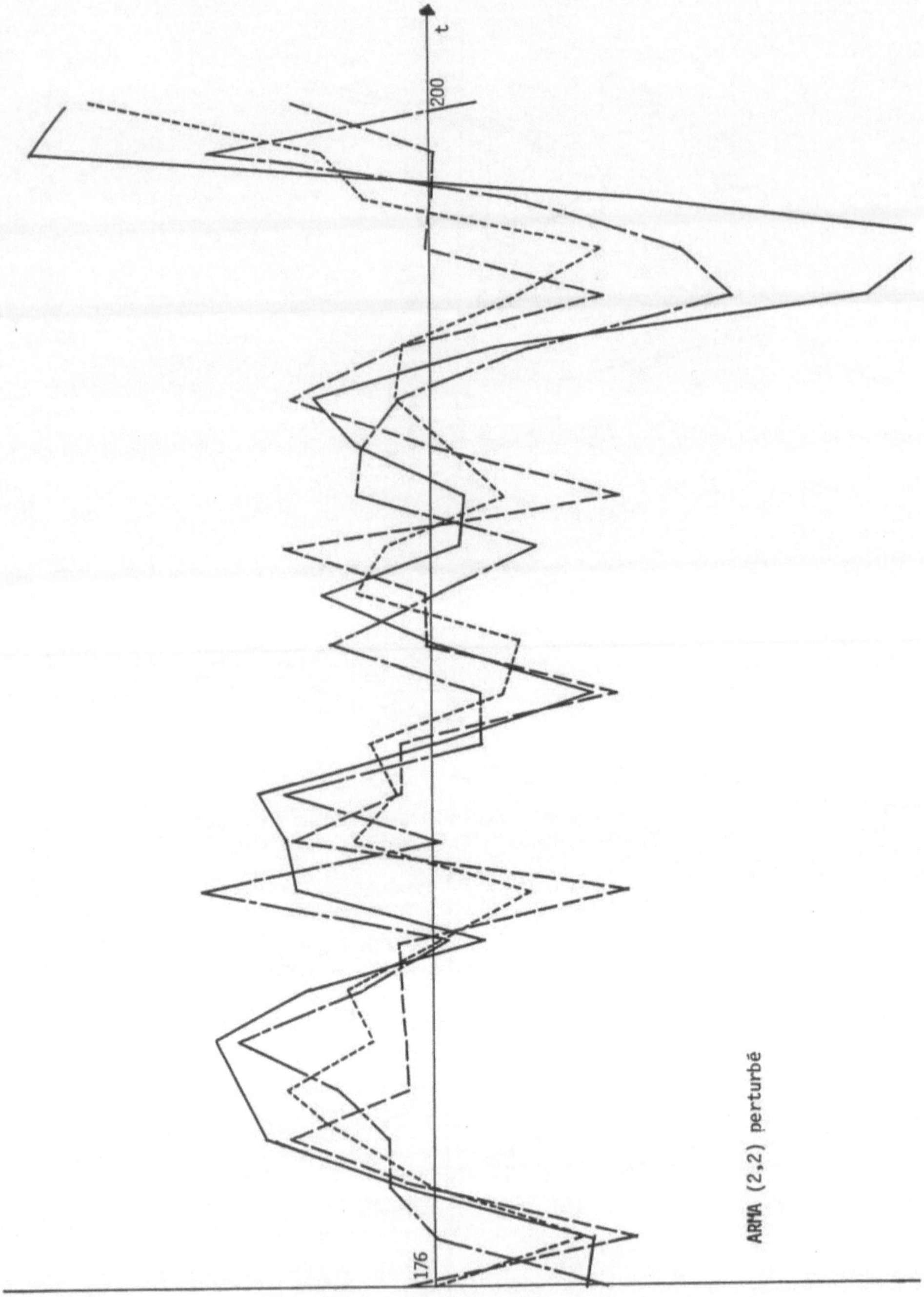
t
200
176
ARMA (2,2) perturbé

APPROXIMATE REDUCTIONS OF BAYESIAN EXPERIMENTS

J.P. Florens
GREQE - EHESS
and
Université d'Aix-Marseille

ABSTRACT

This paper presents some definitions about sufficiency, information value and approximate sufficiency in a bayesian model. The comparison of probabilities through divergences appears to be particularly convenient to define information value and to connect this notion to sufficiency. The properties of two different measures of approximate sufficiency are compared.

Key-words : Sufficiency, divergence, information value, deficiency.

AMS/MOS : Primary 62B05, Secondary 62B10

Acknowledgements : I am grateful to M. Mouchart and J.M. Rolin for helpful comments on an earlier draft of this paper.

1. INTRODUCTION AND NOTATION

A Bayesian experiment is defined by a probability π on a product of measurable spaces $(A \times S,\ \mathcal{A} \otimes \mathcal{S})$. The first space is the parameter space and the second space is the sample space. The marginal probabilities induced by π on $(A,\mathcal{A})$ and $(S,\mathcal{S})$ are called respectively the prior probability and the predictive probability. If $X \in \mathcal{S}$ the conditional probability of X given $\mathcal{A}$ is denoted $P^{\mathcal{A}}(X)$ and is the sampling probability of X. Similarly, if $E \in \mathcal{A}$ the conditional probability of E given $\mathcal{S}$ is denoted $\mu^{\mathcal{S}}(E)$ and is the posterior probability of E. Note that $\mathcal{A}$ and $\mathcal{S}$ represent both the σ-fields on A and S and the σ-fields of cylinders of the product space.

The study of the evolution of the properties of this model when π is replaced by π^* (neighbouring π in some sense) can be considered as an approach to the analysis of robustness in the Bayesian framework. An example in this direction can be found in Stein (1965) who analyses the robustness of Bayesian experiments to a modification of the prior probability.

The situation we want to describe arises when the probability π is replaced by π^* chosen in such a way that a conditional independence condition between sub-σ-fields of $\mathcal{A} \otimes \mathcal{S}$ is satisfied. In our previous work, we characterized admissible reductions of a Bayesian experiment (e.g. sufficiency or ancillarity) by conditional independence conditions; hence the measurement of a proximity criterion between π and π^* will give us a measure of the deficiency of the admissibility of a reduction.

We shall restrict ourselves in this paper to a special case of admissible reduction, the sufficiency of a sub-σ-field of the sampling space, in order to make this intuitive presentation more precise. This case is the most familiar one to statisticians and the extensions to sufficienccy on the parameter space or to the ancillarity do not present any particular problems.

Our paper will be organized as follows. First we shall briefly recall the definition and the main properties if sufficiency. Secondly we shall present some definitions and results concerning the comparison of probabilities essentially based on the concept of divergence. This presentation will involve us in a digression concerning the information value of a Bayesian experiment. Finally we shall examine the notion of approximate sufficiency.

This paper is a synthesis of several previous papers, thus the proofs of the results are not reported here. Details and examples can be found in several papers quoted in the references : Florens (1980), Florens and Mouchart (1979), Florens, Mouchart and Rolin (1980), Florens and Scotto (1982) and Mouchart and Rolin (1979).

2. SUFFICIENCY

Let us consider a sub-σ-field T of S. T will be said to be sufficient if A and S are independent conditionally on T. We denote by $A \perp\!\!\!\perp S \mid T$ this conditional independence.

Several equivalent characterizations of this independence can be given. In terms of sampling probabilities, $A \perp\!\!\!\perp S \mid T$ is equivalent to the a.s. equality between the conditional probabilities of any $X \in S$ given $A \circledast T$ and given T (i.e. $P^{A \circledast T}(X) = P^T(X)$ a.s.). In terms of posterior probabilities this property is equivalent to the a.s. equality between the conditional probabilities of any $E \in A$ given S and given T (i.e. $\mu^S(E) = \mu^T(E)$ a.s.). One can also verify that the sufficiency of T defined by $A \perp\!\!\!\perp S \mid T$ is equivalent to the sufficiency (in the "classic" sense of sampling statistics) of $A \circledast T$ in the classic experiment defined on $(A \times S, A \circledast S)$ by the two probabilities π and $\mu \circledast P$.

We shall say that a Bayesian experiment is dominated if π is dominated by $\mu \circledast P$. In this case the sufficiency of T is equivalent to the existence of $A \circledast T$-measurable derivative of π with respect to $\mu \circledast P$.

3. DIVERGENCES BETWEEN PROBABILITIES AND INFORMATION VALUE

We shall compare probabilities from a different point of view than that of distances. This concept is called φ-divergence and has been studied in particular by Csizar in his important contributions (1967a and b and 1975).

Let us consider a real convex function φ defined on $]0, +\infty[$ and defined at 0 by continuity. φ is usually chosen in such a way that $\varphi(1) = 0$. If (U,u) is a measurable space, let us consider two probabilities λ_1 and λ_2 defined on u. We call φ-divergence between λ_1 and λ_2, denoted by $D\varphi(\lambda_1 \mid \lambda_2)$, the value of the integral $\int_U \varphi\left(\frac{f_1}{f_2}\right) d\lambda_2$ where f_1 and f_2 are respectively the derivatives of λ_1 and λ_2 with respect to a common dominating measure (e.g. $\frac{1}{2}(\lambda_1 + \lambda_2)$) $D\varphi(\lambda_1 \mid \lambda_2)$ does not depend on a particular choice of this dominating measure. Particular choices of φ lead to well-known comparisons of probabilities : if $\varphi(x) = x\text{Log}x$, $D\varphi$ is called the I-divergence or the negative entropy; if $\varphi(x) = \frac{1}{2}|x-1|$, $D\varphi$ is the total variation distance. If $\varphi(x) = (\sqrt{x} - 1)^2$ or $(x-1)^2$, $D\varphi$ is respectively equal to the square of Hellinger distance or to the well-known χ^2.

$D\varphi$ is not, in general, a distance : it is not symmetric and the triangle inequality is not generally verified. However $D\varphi$ satisfies some properties in relation to the decomposition of probabilities into the marginal and conditional probabilities which is the cornerstone of the Bayesian approach. We summarize these properties in the following theorem.

3.1. THEOREM :

a) $\forall\ \lambda_1, \lambda_2$

$$D\varphi(\lambda_1 \mid \lambda_2) \geqslant 0.$$

b) $\forall$ v sub-σ-field of u, we denote λ_{1v} and λ_{2v} the restrictions of λ_1 and λ_2 to v. Then

1° $D\varphi(\lambda_1 \mid \lambda_2) \geqslant D\varphi(\lambda_{1v} \mid \lambda_{2v})$;

2° If for any bounded random variable ξ defined on (U,u) there exists a common determination of $E(\xi \mid v,\lambda_1)$ and $E(\xi \mid v,\lambda_2)$ (the conditional expectations of ξ given v for the two probabilities λ_1 and λ_2), i.e. if v is sufficient - in the classic sense - for λ_1 and λ_2 then

$$D\varphi(\lambda_1 \mid \lambda_2) = D\varphi(\lambda_{1v} \mid \lambda_{2v}).$$

c) If $\lambda_{1v} = \lambda_{2v}$, if there exists a regular conditional probability of λ given v and if one can choose λ_1^v and λ_2^v, the conditional probabilities of λ_1 and λ_2 given v, such that these conditional probabilities are regular and such that $D\varphi(\lambda_1^v \mid \lambda_2^v)$ is v-measurable. Then we get :

$$D\varphi(\lambda_1 \mid \lambda_2) = \int_U D\varphi(\lambda_1^v \mid \lambda_2^v)\, d\lambda_{1v}.$$

d) If φ is *strictly* convex, $D\varphi(\lambda_1 \mid \lambda_2) = 0$ implies $\lambda_1 = \lambda_2$ or, more generally if $D\varphi(\lambda_1 \mid \lambda_2) = D\varphi(\lambda_{1v} \mid \lambda_{2v})$ there exists a common determination of $E(\xi \mid v,\lambda_1)$ and $E(\xi \mid v,\lambda_2)$ for any ξ.

□

A first use of the divergence definition is related to the notion of information value. We define the *information value of a Bayesian experiment* by the φ-divergence between π and $\mu \otimes P$. In the case of regular conditional probabilities one has the relations (see theorem 3.1.d)

$$D\varphi(\pi \mid \mu \otimes P) = \int_A D\varphi(P^A \mid P)\, d\mu = \int_S D\varphi(\mu^S \mid \mu)\, dP.$$

This definition can be found in Lindley (1956) in the case $\varphi(x) = x\mathrm{Log}x$ and in Csizar (1967b) in the general case. Examples are given in Florens and Scotto (1982).

The definition of sufficiency and theorem 3.1. imply the following results.

3.2. THEOREM :

Let T be a sub-σ-field of S and $\pi_{A\otimes T}$ the restriction of π to $A \otimes T$. The information value of the experiment $(A \times S, A \otimes S, \pi)$ is greater than or equal to the information value of the marginal experiment $(A \times S, A \otimes T, \pi_{A\otimes T})$. If T is sufficient these two information values are equal. Reciprocally if φ is *strictly* convex the equality between these quantities implies the sufficiency of T.

□

Let us remark that the total variation distance is not defined by a strictly convex function and the last part of this theorem is false in this case. Counter examples can be easily constructed.

4. APPROXIMATE SUFFICIENCY

There are two possible ways to define the approximate sufficiency of a sub-σ-field T of S.

The first approach was analyzed in particular by Perez (1965) or Csizar (1967b) and we now briefly recall it. T will be called *ε-sufficient* if the difference between the information value of $(A \times S, A \otimes S, \pi)$ and the information value of $(A \times S, A \otimes T, \pi_{A\otimes T})$ is not greater then ε $(\varepsilon > 0)$. It is necessary to assume the strict convexity of φ in order to guarantee good properties of this definition. So we eliminate in particular the total variation distance. The main properties of ε-sufficiency are given by the following theorem.

4.1. THEOREM :

If φ is strictly convex one has

a) : T is sufficient if and only if T is 0-sufficient.

b) : The trivial sub-σ-field of S is ε-sufficient for any ε greater than or equal to the information value of the experiment.

c) : If T' is a sub-σ-field of S such that $T' \subset T$ and if T' is ε-sufficient then T is ε-sufficient. In particular any T is ε-sufficient if ε is greater than or equal to the information value.

d) : If A' and S' are σ-fields such that $A' \subset A$ $T \subset S' \subset S$ and if T is ε-sufficient in $(A \times S, A \otimes S, \pi)$ then T is ε-sufficient in $(A \times S, A' \otimes S'$ $\pi_{A'\otimes S'})$.

□

The second way to consider approximate sufficiency makes the intuitive definition given in introduction more precise. We shall construct a probability π^* on $A \otimes S$ such that T will be sufficient in the experiment defined by π^* and the φ-divergence between π and π^* will be a lower bound of a set of ε such that T is ε-sufficient. More generally, one consider a family of π^* such that $A \perp\!\!\!\perp S \mid T$ for any π^* in this family and takes as this lower bound the quantity $\operatorname{Inf}_{\pi^*} D\varphi(\pi \mid \pi^*)$. For example we can define this family by the set of all probabilities equal to π on $A \otimes T$ and satisfying the conditional independence condition. In this case we essentially get the concept of deficiency proposed by Blackwell (1951) and Lecam (1964).

We want to restrict our attention to a definition obtained with only one probability π^* which is defined by :

$$\forall E \in A,\ X \in S \qquad \pi^* (E \times X) = \int_{E \times S} P^T(X)\ d\pi$$

$P^T(X)$ is a (assumed regular) version of the conditional probability P on S given T. It is easy to check that π^* is well defined and is independent of the choice of a particular version of P^T. One can also verify that T is sufficient in the experiment defined by π^* because π and π^* coincide on $A \circledast T$ and P^T is a.s. a version of the conditional probability π^* on S given $A \circledast T$. Moreover we have the relations :

$$\pi^* (E \times X) = \int_X \mu^T(E)\ dP = \int_E d\mu \int_S P^T(X)\ dP_T^A .$$

All probabilities are assumed regular and P_T^A is the restriction to T of the sampling probability. Then, T is *ε-sufficient* if $D\varphi(\pi \mid \pi^*) \leq \varepsilon$. Let us remark that

$$D\varphi(\pi \mid \pi^*) = \int_S D\varphi(\mu^S \mid \mu^T)\ dP = \int_A D\varphi(P^A \mid P_T^A \circledast P^T)\ d\mu$$

where $P_T^A \circledast P^T$ is a transition probability from (A,A) to (S,S) defined by

$$P_T^A \circledast P^T\ (X) = \int_S P^T(X)\ dP_T^A \qquad \forall X \in S.$$

Properties a), b) and d) of theorem 4.1. are maintained with the preceding definition of ε-sufficiency if $D\varphi(\lambda_1 \mid \lambda_2) = 0$ implies $\lambda_1 = \lambda_2$ (which occurs if φ is strictly convex or if $D\varphi$ is a distance). However we must note that property c) of theorem 4.1. is no longer verified : if $T' \subset T$ is ε-sufficient, it is in general false that T is ε-sufficient. Equivalently, the measure of lack of sufficiency we propose is not monotone relative to the inclusion of sub-σ-field. We have constructed in a previous paper (1980) counter examples to this monotonicity in the cases when $\varphi(x) = \frac{1}{2}|x-1|$, $(\sqrt{x}-1)^2$ or $(x-1)^2$.

The case of the function $\varphi(x) = x\mathrm{Log}x$ is a particular one because the two proposed definitions of ε-sufficiency are equivalent : $D\varphi(\pi \mid \pi^*)$ is exactly the difference between the information values of the experiment and of the marginal experiment restricted to $A \circledast T$.

A particular property occurs if $\varphi(x) = \frac{1}{2}|x-1|$. In this case, the difference between these two information values is less than or equal to any ε for which T is ε-sufficient in the second sense.

In this short paper we only give one example of the uses of ε-sufficiency. Other applications are possible, for example in asymptotic theory or in decision analysis. The example concluding this paper is an approximate version of Basu's theorem (1955 and 1958).

Let us first recall the exact Bayesian version of this result : T_1 and T_2 are sub-σ-fields of S such that T_1 is sufficient and such that T_1 and T_2 are independent (in the predictive probability), then T_2 is ancillary, i.e. A and T_2 are independent. We propose the following extension :

4.2. THEOREM

If T_1 and T_2 are independent (in the predictive probability) and if T_1 is ε-sufficient then T_2 is ε-ancillary in the sense that the information value of the experiment $\{A \times S, A \otimes T, \pi_{A \otimes T}\}$ is less than or equal to ε.

□

5. REFERENCES

Basu, D. (1955) : "On Statistics Independent of a Complete Sufficient Statistic", *Sankhya*, 15, 377-380.

Basu, D. (1958) : "On Statistics Independent of a Sufficient Statistic", *Sankhya*, 20, 223-226.

Blackwell, D. (1951) : "Comparison of Experiments", *Proceedings of the Second Berkeley Symposium in Mathematical Statistics and Probability*, University of California Press, Berkeley.

Csiszar, I. (1967a) : "On Topological Properties of f-divergences", *Studia Sci. math. Hungar.*, 2, 300-318.

Csiszar, I. (1967b) : "On Information Type Measures of Difference of Probability Distributions and Indirect Observations", *Studia Sci. math. Hungar.*, 2, 329-339.

Csiszar, I. (1975) : "I-divergence Geometry of Probability Distributions and Minimization Problems", *Ann. Probab.*, 3, 146-158.

Florens, J.P. (1980) : "Comparison of Marginal Experiments", (revised version), manuscript GREQE, Université d'Aix-Marseille.

Florens, J.P. and M. Mouchart (1979) : "Reduction of Bayesian Experiments" (revised version), CORE D.P. 7737, Université Catholique de Louvain.

Florens, J.P. and M. Mouchart and J.M. Rolin (1980) : "Réduction dans les expériences bayésiennes séquentielles", Colloque Processus Aléatoires et Problèmes de Prévision, Bruxelles 1980, *Cah. Cent. Etud. Rech. Oper.*, 22 (3-4), 353-362.

Florens, J.P. and S. Scotto (1982) : "Information Value and Econometric Modelling", manuscript - GREQE - Université d'Aix Marseille.

Lecam, L. (1964), "Sufficiency and Approximate Sufficiency", *Ann. Math. Statist.*, 35, 1419-1455.

Mouchart, M. and J.M. Rolin (1979) : "A Note on Conditional Independence", Rapport 129, Séminaire de Mathématique Appliquée et Mécanique, Institut de Mathématique Pure et Appliquée, Université Catholique de Louvain.

Perez, A. (1965) : "Information, ε-Sufficiency and Data Reduction Problems", *Kybernetica cislo*, 4, 297-322.

Stein, C. (1965) : "Approximation of Improper Prior Measures by Prior Probability Measures, in *Bernouilli, Bayes, Laplace Anniversary Volume*, ed. by J. Neyman and L. LeCam 217-240, Berlin, *Springer Verlag*.

THEORY AND APPLICATIONS OF LEAST SQUARES APPROXIMATION IN BAYESIAN ANALYSIS

by

M. Mouchart* and L. Simar*,*

Abstract

Least squares approximations of posterior expectations are shown to provide interesting alternatives to exact computations. The theoretical part shows how to take advantage of suitable choices of coordinates and of particular structures of the sampling process. The information extracted from the sample is characterized in terms of the concept of "Least squares sufficiency". Applications to the estimation of a population mean, to prediction problems, to linear models and to the estimation of distribution functions are presented to illustrate the theory and to point out how an approximation to a broader model may offer a useful alternative to the exact solution of a narrower model.

Key-words : Least squares approximations, Linear Bayes methods, Linear inference

AMS/MOS : Primary 62J99, Secondary 62F15

* CORE, Université Catholique de Louvain.

⋆ SMASH, Facultés Universitaires Saint-Louis, Bruxelles.

1. INTRODUCTION

Consider a Bayesian experiment, *i.e.* a probability on the product space formed by a sample space and a parameter space. Let $x \in \mathbb{R}^p$ represent observations and $\theta \in \mathbb{R}^q$ represent parameters. In this paper we assume the Bayesian experiment is such that (x',θ') is square integrable. Let the first two moments be denoted as follows :

$$E\begin{pmatrix}\theta\\ x\end{pmatrix} = \begin{pmatrix}E(\theta)\\ E(x)\end{pmatrix} \tag{1.1}$$

$$V\begin{pmatrix}\theta\\ x\end{pmatrix} = \begin{pmatrix}V_{\theta\theta} & V_{\theta x}\\ V_{x\theta} & V_{xx}\end{pmatrix} \tag{1.2}$$

In this paper we analyze the function $\hat{E}(\theta|x)$ defined as follows :

$$\hat{E}(\theta|x) = E(\theta) + V_{\theta x}V_{xx}^{-1}(x - E(x)). \tag{1.3}$$

This function may be interpreted as a best (in the sense of least squares) approximation of either θ or $E(\theta|x)$ by a linear function of x . Doob (1953) suggested calling $\hat{E}(\theta|x)$ a wide-sense version of $E(\theta|x)$. In order to appreciate the accuracy of the L.S. approximation let us introduce the error term :

$$\eta = \theta - \hat{E}(\theta|x) . \tag{1.4}$$

The variance of this error term is easily shown to be :

$$V(\eta) = V_{\theta\theta} - V_{\theta x}\, V_{xx}^{-1}V_{x\theta}. \tag{1.5}$$

If one is interested in the approximation of $E(\theta|x)$, the error of approximation is given by :

$$E(\eta|x) = E(\theta|x) - \hat{E}(\theta|x). \tag{1.6}$$

Therefore $V(\eta)$ gives an upper bound for the variance of (1.6). Finally, from (1.4), $V(\eta)$ also gives an upper bound for the expected posterior variance of θ ;

$$E\, V(\theta|x) \leqslant V(\eta) \tag{1.7}$$

where $\leqslant$ is written in the sense of positive-definite symmetric (PDS) matrices, and with equality if and only if the true regression is linear (*i.e.*

iff $E(\theta|x) = \hat{E}(\theta|x)$). A more detailed presentation of L.S. approximation in Bayesian analysis, along with more systematic references to the literature, may be found in the first section of Mouchart and Simar (1980).

The object of this paper is to present a survey of various theoretical questions raised by Least Squares Approximation in Bayesian analysis, along with several applications. In comparison with our 1980 paper, the theory has been condensed by not repeating the proofs, however the statistical discussion is somewhat deeper here; on the other hand, more applications have been worked out in order to give more insight into the relevance of these approximations for statistical practice.

The following section considers the problem of choosing a statistic particularly suited for the purpose of L.S. approximation and shows why unbiased estimators lead to attractive forms of these approximations. The third section exhibits, through the concept of L.S. sufficiency, the type of sample information retained by the L.S. approximation. The fourth section analyzes what type of information an enchangeable sampling process supplies for a L.S. approximation. The last section gathers several examples of applications. This involves the estimation of a population mean, the problem of prediction, the analysis of linear models and the estimation of distribution functions.

2. CHOOSING A STATISTIC : THE USE OF UNBIASED ESTIMATORS

We first note that the moments needed for the computation of $\hat{E}(\theta|x)$ in (1.3) may be obtained by integrating sampling moments with respect to the prior probability; more specifically we shall often use the following decomposition :

$$E(x) = E[E(x|\theta)] \tag{2.1}$$

$$\begin{aligned} V(x) = V_{xx} &= V(E(x|\theta)) + E(V(x|\theta)) \\ &= V_b + V_w, \text{ say.} \end{aligned} \tag{2.2}$$

$$\mathrm{cov}(\theta,x') = V_{\theta x} = \mathrm{cov}(\theta, E(x'|\theta)) \tag{2.3}$$

We shall also use the following notation relative to a given partition of θ :

$$\theta' = (\theta_1', \theta_2') \qquad \theta_i \in \mathbb{R}^{q_i} \qquad q_1 + q_2 = q \tag{2.4}$$

$$m^{\circ\prime} = E(\theta') = (m_1^{\circ\prime}, m_2^{\circ\prime}) \tag{2.5}$$

$$V^\circ = V(\theta) = V_{\theta\theta} = \begin{pmatrix} V^\circ_{11} & V^\circ_{12} \\ V^\circ_{21} & V^\circ_{22} \end{pmatrix} \tag{2.6}$$

Motivation

If the joint distribution of (θ,x) were multivariate normal, $\hat{E}(\theta|x)$ and $V(\eta)$ would be equal to the conditional expectation $E(\theta|x)$ and the conditional variance $V(\theta|x)$ respectively. Therefore, one may interpret $\hat{E}(\theta|x)$ and $V(\eta)$ as the conditional expectation and variance of a normal approximation to the actual distribution of (θ,x). This suggests that the quality of the L.S. approximation will crucially depend on the choice of coordinates in the (θ,x)-space, more specifically the choice of a parametrization and the choice of a statistic.

Let us now consider a fixed parametrization θ . Let $t = t(x)$ be a statistic defined on the x-space. The error η_t corresponding to t is defined as :

$$\eta_t = \theta - \hat{E}(\theta|t) . \tag{2.7}$$

Then (1.7) may be generalized as follows :

$$E\ V(\theta|x) \leqslant E\ V(\theta|t) \leqslant V(\eta_t) \qquad \text{a.s. } \forall t = t(x) \tag{2.8}$$

where the equality holds in the first inequality when t is sufficient and holds in the second inequality when $E(\theta|t)$ is linear in t (*i.e.* $\hat{E}(\theta|t) = E(\theta|t)$). One may conclude that a "good" statistic t on which to base the L.S. approximation would be a statistic such that :

1) the sampling moments $E(t|\theta)$ and $V(t|\theta)$ are not too difficult to compute (otherwise the exact solution $E(\theta|x)$ may be preferable);

2) the statistic t is not "too far" from being sufficient and $E(\theta|t)$ is not "too far" from $\hat{E}(\theta|t)$.

In so far as computational simplicity is of basic interest one may not often expect an exact answer to the second issue. Since the interest is in estimating θ (and thus in minimizing $V(\eta_t)$) or in approximating $E(\theta|x)$ (and thus in minimizing $V\ E(\eta_t|x)$, which has $V(\eta_t)$ as an upper bound), it seems natural to look for a statistic leading to a minimum of $V(\eta_t)$. In the sampling theory of point estimation, we know that, in general, no uniformally optimal estimator may be found; this has led to the development of a theory of unbiased estimation. This theory is relatively simple and provides conditions for the existence of (restricted) optimal solutions. This suggests considering the role of unbiased estimators in the L.S. approximation. Furthermore, such estimators will be shown to simplify the computations, to help understand how the prior and the sample information combine and to give more insight into the connection between $\hat{E}(\theta|x)$ and the normal theory.

Use of an unbiased estimator

Suppose the sample information has been concentrated on t , an unbiased estimator of θ_1, *i.e.*

$$E(t|\theta) = \theta_1 \ . \tag{2.9}$$

Here t is a q_1-dimensional statistic. In such a case, we obtain :

$$V_{\theta t} = \text{cov}\ (\theta,\theta_1) = \begin{pmatrix} V^\circ_{11} \\ \\ V^\circ_{21} \end{pmatrix} = \begin{pmatrix} I_{(q_1)} \\ \\ R^\circ_{21} \end{pmatrix} V^\circ_{11} \tag{2.10}$$

where R°_{21} is defined by the last equality. Unbiasedness furthermore implies :

$$V_{b(t)} = V\ E(t|\theta) = V^\circ_{11} \tag{2.11}$$

$$V_{w(t)} = E\ V(t|\theta) \tag{2.12}$$

$$E(t) = m^\circ_1 \tag{2.13}$$

The L.S. approximation of θ becomes :

$$\hat{E}(\theta|t) = m_o + \begin{pmatrix} I_{(q_1)} \\ \\ R^\circ_{21} \end{pmatrix} \left[(V^\circ_{11})^{-1} + V^{-1}_{w(t)}\right]^{-1} V^{-1}_{w(t)}(t - m^\circ_1). \tag{2.14}$$

In particular, $\hat{E}(\theta_1|t)$ has the familiar form of a matrix weighted average :

$$\hat{E}(\theta_1|t) = \left[(V^\circ_{11})^{-1} + V^{-1}_{w(t)}\right]^{-1}\left[(V^\circ_{11})^{-1}\ m^\circ_1 + V^{-1}_{w(t)}t\right] \tag{2.15}$$

$$= (I - A)m^\circ_1 + At$$

where $A = \left[(V^\circ_{11})^{-1} + V^{-1}_{w(t)}\right]^{-1} V^{-1}_{w(t)}$; furthermore :

$$\hat{E}(\theta_2|t) = m^\circ_2 + R^\circ_{21}\ A(t - m^\circ_1). \tag{2.16}$$

As could be expected, the L.S. approximation of θ_2 would not incorporate sample information if θ_1 and θ_2 are, a priori, uncorrelated (*i.e.* $V^\circ_{21} = 0$) .

The average measure of accuracy (1.5) is easily shown to be :

$$V(\eta_t) = V^\circ - \begin{pmatrix} I_{(q_1)} \\ R^\circ_{21} \end{pmatrix} V^\circ_{11}(V^\circ_{11} + V_{w(t)})^{-1} V^\circ_{11} \; [I_{(q_1)} \quad R^\circ_{12}]. \tag{2.17}$$

The role of the sample information may be appreciated by tracing $V_{w(t)}$, the expected sampling variance, in $[V(\eta_t)]^{-1}$ which, by (1.7), is also a lower bound for the (harmonic) mean of the posterior precision :

$$[E(V(\theta|t))]^{-1} \geq (V^\circ)^{-1} + \begin{bmatrix} V^{-1}_{w(t)} & 0 \\ 0 & 0 \end{bmatrix} . \tag{2.18}$$

In (2.18), the sampling improves the lower bound for the elements corresponding to θ_1 only.

If attention focuses on θ_1 , we obtain from (2.15) :

$$V(\eta_1) = \left[(V^\circ_{11})^{-1} + V^{-1}_{w(t)}\right]^{-1} = (I-A)V^\circ_{11}(I-A)' + AV_{w(t)}A' . \tag{2.19}$$

These equalities provide a decomposition of the variance of the error term on θ_1 in terms of the (expected) sampling variance $(V_{w(t)})$ and of the prior variance of θ_1 (V°_{11}). From (2.19) and (1.7) we get a rule of super-additive precision :

$$[E\,V(\theta_1|t)]^{-1} \geq (V^\circ_{11})^{-1} + (V_{w(t)})^{-1} \tag{2.20}$$

with equality if and only if $E(\theta_1|t)$ is linear in t *i.e.* $E(\theta_1|t) = \hat{E}(\theta_1|t)$. The role of the simultaneity in the inference on θ may be appreciated by comparing (2.20) and the block (1,1) in (2.18) : their equality would require $V^\circ_{21} = 0$.

It may be pointed out that formula (2.15) (convex combination of t and m°) and (2.20) (addition of mean precisions in case of equality sign) show that these two characteristic features of the normal theory (*i.e.* normal sampling process with known variance and natural conjugate prior) may be generalized as being a property of processes for which the posterior expectation is linear in an unbiased estimator. This generalizes Ericson (1969).

Instead of considering a fixed parametrization we now consider a situation where the sampling process is kept fixed but robustness w.r.t. the prior specification is of basic concern. Let t again be an unbiased estimator of θ and s be a sufficient statistic. Let t^* be the Rao-Blackwellization of

t w.r.t. s :

$$t^* = t^*(s) = E(t|s,\theta). \qquad (2.21)$$

We shall use superscript $*$ to indicate symbols associated with t^* instead of t . Rao-Blackwell's theorem shows that :

$$V(t^*|\theta) \leq V(t|\theta) \qquad \text{for any } \theta. \qquad (2.22)$$

Therefore

$$V_{w(t^*)} \leq V_{w(t)} \qquad \text{for any prior probability.} \qquad (2.23)$$

Consequently, $\hat{E}(\theta|t^*)$ will improve $\hat{E}(\theta|t)$ in the following sense :

$$V(\eta_{t^*}) \leq V(\eta_t) \qquad \text{for any prior probability.} \qquad (2.24)$$

If, furthermore, t^* is complete, Lehman-Scheffé's theorem implies that t^* is UMVUE and therefore $V(t^*|\theta)$ is minimal for any θ : consequently, $V_{w(t^*)}$ and $V(\eta_{t^*})$ are minimal , among all unbiased estimators , for any prior distribution.

3. THE USE OF SAMPLE INFORMATION : L.S. SUFFICIENCY

What do we learn from the data when using L.S. approximation ? Clearly, we only retain the linear function (1.3). A statistic $t = t(x)$ would not lose information (for the L.S. approximation) if $\hat{E}(\theta|x) = \hat{E}(\theta|t(x))$ a.s.; as the l.h.s. is a linear function of x and the r.h.s. is a linear function of $t(x)$, such an equality can be expected only if $t(x)$ is linear in x . This motivates the following concept of L.S. sufficiency :

THEOREM. (definition of L.S. sufficiency)

Let $t(x) = Ax \qquad A : s \times p$

then the following conditions are equivalent and define $t(x)$ *as L.S. sufficient for* θ *(relative to* x *) :*

(i) $\hat{E}(\theta|x) = \hat{E}(\theta|t(x))$ x*.-a.s.*

(ii) $C(V_{xx}^{-1}V_{x\theta}) \subseteq C(A')$ *where* $C(\cdot)$ *is the linear space generated by the columns of a matrix ;*

(iii) $\exists B : q \times s$ *such that* $V_{\theta x} V_{xx}^{-1} = B A$.

DEFINITION. $t(x) = Ax$ *is minimal L.S. sufficiency for* θ *iff*

$$C(A') = C(V_{xx}^{-1} V_{x\theta})$$

In other words, a minimal L.S. sufficient statistic may be constructed from any basis of $C(V_{xx}^{-1} V_{x\theta})$. For a proof of the theorem and more details see Mouchart and Simar (1980). Consider now :

$$t^* = V_{\theta x} V_{xx}^{-1} x \tag{3.1}$$

Then t^* is not only a L.S. sufficient statistic, it is also a "best" linear statistic for the L.S. approximation in the sense that for any linear $t = Ax$:

$$V(\eta_{t^*}) \leq V(\eta_t) , \tag{3.2}$$

with equality if and only if t is also L.S. sufficient.

4. EXCHANGEABILITY

We now consider exchangeable processes, *i.e.* processes where the finite dimensional distributions are invariant under permutation of indices (see *e.g.* Hewitt and Savage (1958) or Chow and Teicher (1978, chap. 7)). This class is stable for mixtures; in particular, it includes the mixtures of I.I.D. processes; therefore such processes naturally arise *e.g.* after integrating out nuisance parameters in I.I.D. processes.

First order exchangeable processes form the (very large) class of processes which verify the condition of exchangeability only for the first moment *i.e.* under suitable reparametrization, they are such that for any sample $x = (x_1, \ldots, x_n)$:

$$E(x|\theta) = \theta_1 e \qquad \text{where} \qquad e' = (1 \ldots 1) \in \mathbb{R}^n \tag{4.1}$$

and θ_1 is the first component of θ. This implies that $V_{\theta x} = v_1^\circ e'$ where v_1° is the first column of $V^\circ = V(\theta)$. It has been shown by formula (2.16) in Mouchart and Simar (1980) that this implies : $C(V_{xx}^{-1} V_{x\theta}) = C(V_w^{-1} e)$; consequently the *one-dimensional* statistic :

$$t^+(x) = \frac{e' V_w^{-1} x}{e' V_w^{-1} e} \tag{4.2}$$

is minimal L.S. sufficient for the complete vector θ and unbiased for θ_1. Therefore the results of Section 2 apply. Since :

$$\text{E} V(t^{+}|\theta) = [e'V_w^{-1}e]^{-1} \underset{\text{def}}{=} v_{w(t^{+})} \,, \tag{4.3}$$

the formulae (2.14) to (2.20) are still valid when V_w is replaced by $v_{w(t^+)}$. In particular :

$$\hat{E}(\theta|x) = \hat{E}(\theta|t^{+}) = m_o + \frac{t^{+} - m_1^{\circ}}{V(t^{+})} v_1^{\circ} \tag{4.4}$$

where

$$V(t^{+}) = v_{11}^{\circ} + v_{w(t^{+})}. \tag{4.5}$$

The rule of additive precision (2.18) now becomes :

$$\left[E\, V(\theta|t^{+})\right]^{-1} \geq V_o^{-1} + (v_{w(t^{+})})^{-1} \begin{bmatrix} 1 & 0 & \dots & 0 \\ 0 & & & \vdots \\ 0 & & & \vdots \\ 0 & \dots & \dots & 0 \end{bmatrix} \tag{4.6}$$

Second order exchangeable processes are such that V_w has an intraclass correlation structure; therefore e is an eigenvector V_w and $t^{+}(x)$ is the sample mean, $\bar{x}$. Note that, in time series analysis, first-order exchangeability is implied by stationarity while second-order exchangeability is not.

5. APPLICATIONS

5.1. Estimation of a Population Mean

Consider an n-sample $x = (x_1,\dots,x_n)'$ drawn from a first-order exchangeable process :

$$E(x|\theta) = \theta_1\, e \qquad e = (1,\dots,1)' \in \mathbb{R}^n \tag{5.1}$$

and where θ_1 is the first component of θ . Suppose θ_1 is the only parameter of interest. First-order exchangeability implies that the real-valued statistic t^{+} :

$$t^{+}(x) \underset{\text{def}}{=} e'V_w^{-1}x\, [e'V_w^{-1}e]^{-1} \tag{5.2}$$

is both (minimal) L.S. sufficient and unbiased for θ_1. From Sections 2 and 4 we conclude that :

$$\hat{E}(\theta_1|x) = \hat{E}(\theta_1|t^{+}) = (1-a)m_1^{\circ} + at^{+} \tag{5.3}$$

where

$$a = \frac{(v_{w(t^+)})^{-1}}{(v^\circ_{11})^{-1} + (v_{w(t^+)})^{-1}} = \frac{v^\circ_{11}}{v^\circ_{11} + v_{w(t^+)}}$$

The average measure of accuracy is now, from (2.18), :

$$V(\eta_{t^+}) = \left[(v^\circ_{11})^{-1} + (v_{w(t^+)})^{-1}\right]^{-1} \tag{5.4}$$

The inequality (1.7) may now be strengthened as follows :

PROPOSITION. In first-order exchangeable processes one has :

$$E\ V(\theta_1|x) = \left[(v^\circ_{11})^{-1} + (v_{w(t^+)})^{-1}\right]^{-1} \tag{5.5}$$

if and only if :

$$E(\theta_1|x) = \hat{E}(\theta_1|t^+). \tag{5.7}$$

Indeed, in the following inequalities :

$$[E\ V(\theta_1|x)]^{-1} \geqslant [E\ V(\theta_1|t^+)]^{-1} \geqslant (v^\circ_{11})^{-1} + (v_{w(t^+)})^{-1}\ , \tag{5.7}$$

the first one is an equality if and only if $E(\theta_1|x) = E(\theta_1|t^+)$ - *i.e.* t^+ is "sufficient in the mean" - and the second one is an equality if and only if $E(\theta_1|t^+) = \hat{E}(\theta_1|t^+)$ - *i.e.* linear posterior expectation of θ_1 w.r.t. t^+.

□

REMARK 1. Suppose we look for a linear function of x :

$$t(x) = c'x + b\ ; \tag{5.8}$$

that is an unbiased estimator of θ_1 where $E(x|\theta) = \theta_1 e$. This implies $b = 0$ and $c'e = 1$. Then, t^+ defined in (5.2), is optimal in any of the following senses :

$$t^+ = \arg\min_t V(\eta_t) = \arg\min_t V(t) = \arg\min_t E\ V(t|\theta) \tag{5.9}$$

where the minimum is taken with respect to the linear unbiased estimators of θ ; in words : t^+ is the statistic leading to the best (in the sense of $V(\eta_t)$) Least Squares approximation but, in itself, it is also the linear unbiased estimator of θ_1 which minimizes both its predictive variance $(V(t))$ and its *expected* sampling variance. This last aspect may be viewed as a Bayesian analogue to the Gauss-Markov Theorem.

REMARK 2. Suppose that, in first-order exchangeable processes, one uses $\bar{x}$ instead of t^+. Clearly $\bar{x}$ is still unbiased for θ_1. Therefore, from Section 2, one obtains :

$$\hat{E}(\theta_1|\bar{x}) = (1-b)m_1^\circ + b\,\bar{x} \tag{5.10}$$

where

$$b = \frac{v_{11}^\circ}{v_{11}^\circ + v_{w(\bar{x})}} \tag{5.11}$$

$$v_{w(\bar{x})} = E\,V(\bar{x}|\theta) = \frac{e'V_w e}{n^2}\,. \tag{5.12}$$

In general, one has :

$$V(\eta_{t^+}) \leq V(\eta_{\bar{x}}) = \left[(v_{11}^\circ)^{-1} + (v_{w(\bar{x})})^{-1}\right]^{-1} \tag{5.13}$$

with equality if and only if $\bar{x}$ is L.S. sufficient or, equivalently, if and only if e is an eigenvector of V_w ; indeed inequality (5.13) is equivalently written as : $(e'e)^2 \leq (e'V_w e)(e'V_w^{-1}e)$. Such is the case in second-order exchangeable processes where $v_{w(\bar{x})}$ may be written explicitly as :

$$v_{w(\bar{x})} = n^{-1}(m_2 + (n-1)m_3) \tag{5.14}$$

where $m_2 = E\,V(x_i|\theta)$ and $m_3 = E\,\mathrm{cov}(x_i, x_j|\theta)$ $(i \neq j)$. Note that uncorrelatedness in the sampling (*i.e.* $\mathrm{cov}(x_i, x_j|\theta) = 0$) does not provide substantial simplification. Note also that in finite samples, negatively correlated observations (which implies $m_3 < 0$) improve the quality of the L.S. approximation (in comparison with independent sampling) but, in large samples, m_3 may not be negative (because of the restriction $-V(x_i|\theta)(n-1)^{-1} \leq \mathrm{cov}(x_i, x_j|\theta) \leq V(x_i|\theta)$).

Ericson's (1969) result may be written as follows : "In first order exchangeable processes, if $E(\theta_1|x) = \hat{E}(\theta_1|\bar{x})$, then $E(\theta_1|\bar{x})$ has the form (5.10) (5.11)". This may be viewed as a corollary of the above proposition (formulae (5.5) - (5.6) once it is realized that $E(\theta_1|x) = \hat{E}(\theta_1|\bar{x})$ implies that $\bar{x} = t^+$. In his (1970) paper, Ericson introduces second order exchangeability in the context of sampling from finite populations.

5.2. Prediction

We now consider $x' = (y', z')$ where y represents future observations while z represents past data. In this section we specify : $y \in \mathbb{R}^p$, $z \in \mathbb{R}^n$ and, consequently, $x \in \mathbb{R}^{n+p}$ and still $\theta \in \mathbb{R}^q$. The problem of L.S. approximation of a (Bayesian) prediction is to compute and to analyze $\hat{E}(y|z)$.

In general, one has :

$$\hat{E}(y|z) = E(y) + V_{yz} V_{zz}^{-1} (z - E(z)) \tag{5.15}$$

where the predictive moments are computed through (2.1) and (2.2). The use of particular structures such as unbiased statistics (y or z or both) or first order exchangeability does not provide substantial simplification of (5.15) unless either $\operatorname{cov}(y,z|\theta)$ or $\operatorname{cov}(y,z)$ has a particular form. This may be obtained through second order exchangeable processes. Note that exchangeability in the sampling process implies exchangeability in the predictive process.

We therefore analyze situations where :

$$E(x) = E(y',z')' = m_1^\circ (e'_{(n)}, e'_{(p)})' \tag{5.16}$$

$$V(x) = V(y',z') = (a-b)\, I_{(n+p)} + b\, e_{(n+p)} e'_{(n+p)} \tag{5.17}$$

where $e_{(k)} = (1 \ 1 \ \dots \ 1)' \in \mathbb{R}^k$ and $-(n+p-1)^{-1}\, a \leqslant b \leqslant a$.

With the structure (5.16) - (5.17) , we have :

$$V_{yz} V_{zz}^{-1} = b(a + (n-1)b)^{-1} e_{(p)}\, e'_{(n)} \ . \tag{5.18}$$

From Section 3, we again conclude that the sample mean $\bar{z}$ is L.S. sufficient for the prediction; more specifically :

$$\hat{E}(y|z) = \hat{E}(y|\bar{z}) = e_p\{(1-c)m_1^\circ + c\bar{z}\} \tag{5.19}$$

where $c = b\, n(a + (n-1)b)^{-1}$. Let η be the prediction error :

$$\eta = y - \hat{E}(y|\bar{z}) \ . \tag{5.20}$$

We have, from (1.5) and (1.7), :

$$E\, V(y|\bar{z}) \leqslant V(\eta) = (a-b)\, I_{(p)} + e_{(p)} e'_{(p)}\, b(a-b)(a + (n-1)b)^{-1} \ . \tag{5.21}$$

In particular :

$$V(\eta_i) = (a-b)(a+bn)(a+(n-1)b)^{-1} \qquad 1 \leqslant i \leqslant p \tag{5.22}$$

which tends to $a-b$ when $n \to \infty$.

<u>*REMARK*</u> . If exchangeability comes from the sampling process, one has

$$a = m_2 + v_{11}^\circ \tag{5.23}$$

$$b = m_3 + v^\circ_{11} \tag{5.24}$$

where $m_2 = E\,V(x_i|\theta)$, $m_3 = E\,\mathrm{cov}\,(x_i, x_j|\theta)$ and $v^\circ_{11} = V\,E(x_i|\theta)$.

For simplification, let us suppose $p = 1$. It is interesting to compare (5.10) (with $\bar{x}$ replaced by $\bar{z}$) and (5.19); both have the form of a weighted average of m°_1 and $\bar{z}$ with weights inversely proportional to $V(\bar{z}) = n^{-1}(a + (n-1)b)$; this involves :

$$\hat{E}(y|\bar{z}) = \hat{E}(\theta_1|\bar{z}) + m_3(\bar{z} - m^\circ_1)(V(\bar{z}))^{-1} \quad . \tag{5.25}$$

Note that in (5.25) the "correction" through $\bar{z} - m^\circ_1$ disappears when $m_3 = 0$.

5.3. Linear Models

Consider the following linear model :

$$E(x|\theta) = Z\,\beta \qquad Z : n \times k \quad \text{(non random)} \tag{5.26}$$

$$V(x|\theta) = \Omega(\alpha) \tag{5.27}$$

with parameters $\theta' = (\alpha', \beta') \in \Theta \subset \mathbb{R}^{r+k}$. The statistic t^+ in Section 4 may be generalized as follows :

$$b^+ = (Z'\,V_w^{-1}\,Z)^{-1}\,Z'\,V_w^{-1}\,x \tag{5.28}$$

where $V_w = E\,\Omega(\alpha) = E\,V(x|\theta)$; it has been shown in Mouchart and Simar (1982) that b^+ is L.S. sufficient for θ and unbiased for β . While $\hat{E}(\beta|b^+)$ may be a reasonable approximation, it is nevertheless questionable to estimate α through linear functions of x (or of b^+). Clearly, a good choice of coordinates should depend on the structure of $\Omega(\alpha)$ and no general results may be expected. In the case of spherical residuals, *i.e.* :

$$\Omega(\alpha) = \alpha\,I_{(n)} \qquad \alpha > 0 \tag{5.29}$$

b^+ becomes the ordinary Least Squares estimator of β $(b^+ = b = (Z'Z)^{-1}Z'x)$ and a natural choice of unbiased estimator of θ would be $t = (b, s^2)$ where $s^2 = (n-k)^{-1}x'\,M\,x$ with $M = I - Z(Z'Z)^{-1}Z'$. In such a case, $\hat{E}(\theta|b, s^2)$ may be computed as in Section 2 - formula (2.15). In particular, we have :

$$V_{w(t)} = E\,V(b, s^2|\theta)$$
$$= \begin{bmatrix} (n-k)^{-2}\{(m_4 - 3A_o)d'_M d_M + 2(n-k)A_o\} & (n-k)^{-1}m_3(Z'Z)^{-1}Z'd_M \\ (n-k)^{-1}\,m_3\,d'_M Z(Z'Z)^{-1} & a_o(Z'Z)^{-1} \end{bmatrix} \tag{5.30}$$

where $a_o = E(\alpha)$, $A_o = E(\alpha^2)$, $m_j = E(x_i - z_i'\beta)^j$ $j = 3,4$ and d_M is a vector whose elements are the diagonal elements of M . In (5.30), the terms in m_3 and m_4 allow to take into account non-normality in the residuals at a low computational cost; this suggests that, in some circumstances, it may be more appropriate to use a L.S. approximation for a non-normal model than an exact-solution to a model where the conditional distribution of $(x|Z)$ is approximated by a normal distribution. In other words, this example shows that the problem should be viewed not as whether to approximate or not but rather as how to monitor the approximations underlying any statistical model (further analysis of L.S. approximation in linear models may be found in Mouchart-Simar (1982) , see also Goldstein (1976)).

5.4. Estimation of a Distribution Function

An important problem, in non-parametric statistics, is the estimation of the (cumulative) distribution function F_x . If one considers a fixed point x_o , one may define a function θ of the parameter F_x as :

$$\theta = F_x(x_o) = P(x \leq x_o | F_x). \tag{5.31}$$

Given an I.I.D. sample $x = (x_1,\ldots,x_n)$, a natural unbiased estimator of θ is given by the empirical distribution function Z_n at x_o :

$$Z_n = n^{-1} \sum_{i=n}^{n} y_i \qquad y_i = 1_{(-\infty,\, x_o]}(x_i) \qquad 1 \leq i \leq n. \tag{5.32}$$

From Section 2, we obtain :

$$\hat{E}(\theta|Z_n) = a\, m_1 + (1-a) Z_n \tag{5.33}$$

where :
$$a = A\,(A+n)^{-1}$$
$$A = v_{11}^{-1}\,(m_1\,(1-m_1) - v_{11})$$
$$m_1 = E(\theta)$$
$$v_{11} = V(\theta)\ .$$

A natural question is whether solution (5.33) is consistent in the sense of the stochastic process defined by F_x *i.e.* whether there exists a prior distribution on F_x that gives (5.33) as a L.S. approximation for any x_o . That it is so may be viewed by remarking that (5.33) is the exact solution under a Dirichlet prior distribution on F_x ; in such a case, a may also be written as $a = \alpha(\mathbb{R})\,(\alpha(\mathbb{R}) + n)^{-1}$ where $\alpha(\cdot)$ is the parameter of the Dirichlet prior probability.

More details on this topic may be found in Goldstein (1975 a,b), Ferguson (1974) and Rolin (1982).

REFERENCES

Chow, Y-s. and H. Teicher (1978), *Probability Theory, Independence, Interchangeability, Martingales*. New York : Springer-Verlag.

Doob, J.L. (1953), *Stochastic Processes*. New York : John Wiley.

Ericson, W.A. (1969), "A Note on the Posterior Mean of a Population Mean", *Journal of the Royal Statistical Society, Series B*, 31, 332-334.

Ericson, W.A. (1970), "On the Posterior Mean and Variance of a Population Mean" *Journal of the American Statistical Association 65*, 649-652.

Ferguson, T.S. (1974), "Prior Distributions on Spaces of Probability Measures", *Annals of Statistics 2*(4), 615-629.

Goldstein, M. (1975a), "Approximate Bayes Solutions to Some Non-parametric Problems", *Annals of Statistics 3*, 512-517.

Goldstein, M. (1975b), "A Note on Some Bayesian Non-parametric Estimates", *Annals of Statistics 3*, 726-740.

Goldstein, M. (1976), "Bayesian Analysis of Regression Problems", *Biometrika 63*(1), 51-58.

Hewitt, E. and L. O. Savage (1955), "Symmetric Measures on Cartesian Products", *Transactions of the American Mathematical Society 80*, 470-501.

Mouchart, M. and L. Simar (1980), "Least Squares Approximation in Bayesian Analysis", in : *Bayesian Statistics*, Proceedings of the First International Meeting held in Valencia (Spain), May2 - June 2, 1979, Valencia : University Press.

Mouchart, M. and L. Simar (1982), "A Note on Least-Squares Approximation in the Bayesian Analysis of Regression Models", CORE Discussion Paper 8233, Université Catholique de Louvain, Louvain-la-Neuve, Belgium.

Rolin, J.-M. (1982), "Non Parametric Bayesian Statistics : A Stochastic Process Approach", CORE Discussion Paper 8225, Université Catholique de Louvain, Louvain-la-Neuve, Belgium, in this volume, chapter 8.

Non Parametric Bayesian Statistics :

A Stochastic Process Approach

by

J.M. Rolin*

May 1982

Abstract

The specification of a prior distribution on the set of all distribution functions permits to consider a non parametric bayesian experiment as an abstract probability space on which are defined the sampling process and a stochastic distribution process.

In this framework, the Dirichlet process may be characterized, in several ways, in terms of one-dimensional distributions and independence relations between associated σ-algebras. These independence relations naturally define extensions of the Dirichlet process called neutral processes. The power of this approach may be appreciated in three ways. It significantly simplifies the computation of the posterior distribution, it gives new characterizations of the Dirichlet process, and finally it solves Docksum's conjecture that the only distribution process that is both neutral to the right and neutral to the left is the Dirichlet process.

Key words : Stochastic distribution processes, Dirichlet processes, Neutral processes, Conditional independence, Increasing processes with independent increments.

AMS - MOS : Primary 62G99, Secondary 62A15, 60G10.

* PROB and CORE, University of Louvain.

1. INTRODUCTION

The general problem of statistical analysis is to make inference about the sampling distribution of the observations. Most of the time, the statistician uses quite restrictive assumptions about the family of possible distributions for his data. This has the advantage of simplifying the model under consideration and of facilitating the computation of optimal decisions. But two problems are linked with such a procedure. Firstly, how to be sure that the real sampling distribution is in the restricted family of distributions resulting from these assumptions ? Secondly, regarding the decisions resulting from the model, are they strongly dependent on the assumptions made ? A way of dealing with these questions is provided by non parametric statistical analysis which considers a very large family of distributions for the observations and, in the limit, the family of *all* distributions.

From a bayesian point of view, the problem is how to define on this "parameter space", *i.e.* the set of all sampling distributions, a σ-algebra of subsets and how to specify on this measurable space a prior distribution. Having done this, in the cross product space the sampling distribution will become a random probability measure. Three problems arise in the specification of such a prior distribution :

I) Is the set of generated sampling distributions large enough ?

II) How may a prior distribution reflect the prior knowledge of the statistician ?

III) What are the prior distributions which lead to tractable analytical insights into the posterior distribution.

From a robust point of view, in every neighbourhood of the true sampling distribution, there must be a realization of the sampling distribution generated by the prior distribution. Therefore, the corresponding support of the prior distribution has to be very large. This paper may be considered as an attempt to clarify these questions.

2. FORMULATION

In this paper, we will consider that the observation is a simple random sample of a real random variable. So, if R is the real line and $\mathcal{B}$ the borel subsets of R, there exist a measurable space (Ω_0, A_0) and measurable functions $X_k : (\Omega_0, A_0) \to (R, \mathcal{B})$, $1 \leq k \leq n$, such that $X = \{X_k : 1 \leq k \leq n\}$ represents the observation. Moreover to each distribution function F, there corresponds a probability measure P_F on (Ω_0, A_0) such that

$$P_F\left[\bigcap_{1 \le k \le n} \{X_k \leq t_k\}\right] = \prod_{1 \le k \le n} F(t_k) \tag{2.1}$$

One way to define a prior distribution on F, is to provide for each n, for each set $\{t_k : 1 \leq k \leq n\}$ such that $-\infty = t_0 < t_1 < t_2 \ldots < t_n < t_{n+1} = +\infty$ a probability measure $\lambda_{t_1,t_2,\ldots t_n}$ on the set $\{x \in R^n : 0 < x_1 < x_2 < \ldots < x_n < 1\}$, that represents the distribution of $\{F(t_k) : 1 \leq k \leq n\}$ and forms a projective system, *i.e.*, for each ℓ the distribution of $\{F(t_k : 1 \leq k \leq n, k \neq \ell\}$ is the marginal of the distribution of $\{F(t_k) : 1 \leq k \leq n\}$. Equivalently, one may provide a probability measure $\nu_{t_1,t_2,\ldots,t_n}$ on the set $[0,1]^n$ representing the distribution of $\{F(t_k) - F(t_{k-1}) : 1 \leq k \leq n\}$ and having the property that, for each ℓ, the distribution of $\{F(t_1), \ldots, F(t_{\ell+1}) - F(t_{\ell-1}), \ldots, F(t_n) - F(t_{n-1})\}$, *i.e.* $\nu_{t_1,\ldots,t_{\ell-1},t_{\ell+1},\ldots t_n}$, is the same as the distribution of $\{F(t_1), \ldots, F(t_{\ell+1}) - F(t_\ell) + F(t_\ell) - F(t_{\ell-1}), \ldots, F(t_n) - F(t_{n-1})\}$ computed from $\nu_{t_1,\ldots,t_{\ell-1},t_\ell,t_{\ell+1},\ldots,t_n}$. If this projective system is such that

$$\mu(t) = \int_{-\infty}^{+\infty} \tau \, \lambda_t(d\tau) = E[F(t)] \tag{2.2}$$

is a distribution function, then using the Kolmogorov extension theorem, it can be shown (Ferguson 1973, Docksum, 1974) that there exist a probability space (Ω_1, A_1, P_1) and a distribution process $F = \{F_t : -\infty < t < +\infty\}$, *i.e.*

(i) $\forall t$ F_t is a random variable

(ii) for all ω_1, the map $t \to F_t(\omega_1)$ is a distribution function having the given probability measures as finite marginals.

Next we define, the product probability space

$$\begin{aligned} \Omega &= \Omega_1 \times \Omega_0 \\ A &= A_1 \otimes A_0 \\ P &= P_1 \cdot P_F \end{aligned} \tag{2.3}$$

i.e. $\forall B \in A_1$, $\forall A \in A_0$

$$P(B \times A) = E_1[P_F(A) \, 1_B]$$

where E_1 is the expectation with respect to P_1.

We also define, with a slight abuse of notation, for $\omega = (\omega_1, \omega_0)$, the coordonate maps,

$$F_t(\omega) = F_t(\omega_1)$$

$$X_k(\omega) = X_k(\omega_0).$$

As usual, in the general theory of stochastic processes, we define on (Ω, A) the σ-algebras related to the distribution process F, as follows

$$F_s^t = \sigma\{F_u : s < u \leq t\} \quad \forall -\infty \leq s < t \leq +\infty \tag{2.4}$$

Note that F_s^t is a σ-algebra of cylinder sets with base in Ω_1. F_s^t is in fact the information contained in the knowledge of the distribution process on the interval $(s,t]$.

In this cross-product space, the properties of the random sample become : conditionnally on the distribution process, the random variables $\{X_k : 1 \leq k \leq n\}$ are independent, which we write

$$\underset{1 \leq k \leq n}{\perp\!\!\!\perp} X_k \mid F_{-\infty}^{+\infty} \tag{2.5}$$

and are identically distributed with F as distribution function, *i.e.*, for each bounded Borel function f on (R,B),

$$E[f(X_k) \mid F_{-\infty}^{+\infty}] = \int_{-\infty}^{+\infty} f(t)\, dF_t \tag{2.6}$$

and this implies

$$E[f(X_k)] = \int_{-\infty}^{+\infty} f(t)\, d\mu(t).$$

The second member of (2.6) is a Stieltjes (stochastic) integral and has to be understood as follows

$$\int_{t_1}^{t_2} dF_t(\omega) = F_{t_2}(\omega) - F_{t_1}(\omega)$$

The problem of finding the posterior distribution, is to compute $E(Y \mid X)$ for each bounded $F_{-\infty}^{+\infty}$ - measurable random variable Y. Along these lines, a most tractable prior distribution is the Dirichlet process as introduced by Ferguson (1973). This will be the subject of the next section.

3. THE DIRICHLET PROCESS

In this section, we will not only review the well-known definitions and properties of the Dirichlet process (Ferguson 1973 and 1974, Docksum 1974), but we will also write these properties, as in the theory of stochastic processes, in terms of independence relations between associated σ-algebras and in terms of one-dimensional distributions. This new approach leads to significant simplifications of the proofs of the main properties and shows easily how to genaralize the Dirichlet process. It also leads to new characterizations of this process in terms of independence pretty much the same way as, the characterizations of the Dirichlet distribution in the finite case (see Fabius 1973, James and Mosimann 1980).

Let $\alpha : (-\infty,+\infty) \to [0,+\infty)$ be an increasing right continuous function with the property that $\lim_{t\to-\infty} \alpha(t) = 0$ and $\lim_{t\to+\infty} \alpha(t) = \alpha(\infty) < \infty$ then we say that the distribution process F is a Dirichlet process with parameter α and we write

$$F \sim Di(\alpha) \tag{3.1}$$

if and only if, for each n, for each set $\{t_k : 1 \leq k \leq n\}$ such that $-\infty = t_0 < t_1 < t_2 \ldots < t_n < t_{n+1} = +\infty$ the random vector $\{F_{t_k} - F_{t_{k-1}} : 1 \leq k \leq n+1\}$ has a Dirichlet distribution with parameter $\{\beta_k : 1 \leq k \leq n+1\}$ where $\beta_k = \alpha(t_k) - \alpha(t_{k-1})$, *i.e.*

$$P\left[\bigcap_{1\leq k\leq n} \left\{F_{t_k} - F_{t_{k-1}} \in dx_k\right\}\right] = p(x_1,\ldots,x_n)\, dx_1\, dx_2 \ldots dx_n \tag{3.2}$$

where

$$p(x_1,\ldots,x_n) = \Gamma[\alpha(\infty)] \frac{\left(1 - \sum_{1\leq k\leq n} x_k\right)^{\beta_{n+1}-1}}{\Gamma(\beta_{n+1})} \prod_{1\leq k\leq n} \frac{x_k^{\beta_k-1}}{\Gamma(\beta_k)}$$

$$\text{on } \{x_1 > 0, \ldots, x_n > 0, \sum_{1\leq k\leq n} x_k < 1\}$$

$$= 0 \text{ elsewhere.}$$

Equivalently

$$P\left[\bigcap_{1\leq k\leq n} \left\{F_{t_k} \in dy_k\right\}\right] = q(y_1,\ldots,y_n)\, dy_1 \ldots dy_n \tag{3.3}$$

where

$$q(y_1,\ldots,y_n) = \Gamma[\alpha(\infty)] \frac{y_1^{\beta_1-1}}{\Gamma(\beta_1)} \frac{(1-y_n)^{\beta_{n+1}-1}}{\Gamma(\beta_{n+1})} \prod_{2\leq k\leq n} \frac{(y_k-y_{k-1})^{\beta_k-1}}{\Gamma(\beta_k)}$$

$$\text{on } \{0 < y_1 < y_2 \ldots < y_n < 1\}$$

$$= 0 \text{ elsewhere.}$$

It is easy to see that this is a projective system of probability measures. Moreover, since for each $-\infty \leq s < t \leq +\infty$, $F_t - F_s \sim B[\alpha(t) - \alpha(s), \alpha(\infty) - \alpha(t) + \alpha(s)]$ where $B(a,b)$ denotes the Beta distribution, we have

$$\mu(t) = \frac{\alpha(t)}{\alpha(\infty)} . \tag{3.4}$$

Hence, the conditions of the existence of such a process are satisfied.

In view of (3.4), we see that the parameter of the Dirichlet process is proportional to the prior expected value of the distribution function, since

$$\alpha(t) = \alpha(\infty)\ E(F_t) = \alpha(\infty)\ P(X \leq t). \tag{3.5}$$

The proportionality constant $\alpha(\infty)$ may be viewed as a variance factor since

$$\begin{aligned} V(F_t) &= \frac{\alpha(t)\ \{\alpha(\infty) - \alpha(t)\}}{\alpha(\infty)^2\ \{\alpha(\infty) + 1\}} \\ &= \frac{\mu(t)\ \{1 - \mu(t)\}}{\alpha(\infty) + 1} . \end{aligned} \tag{3.6}$$

It follows from this that the Dirichlet process may "easily" reflect the prior knowledge of the statistician.

Moreover, if we use the topology of pointwise convergence, it was shown (Ferguson 1973) that a distribution function G is in the support of the Dirichlet process if and only if G is absolutely continuous with respect to α. In the topology of vague convergence, G is in the support of the Dirichlet process if and only if the support of G is included in the support of α. And so if the support of α is R, every distribution function is in the vague support of the Dirichlet process. (Remember that G is in the support of a distribution process F if and only if the probability that the distribution process is in a neighbourhood of G is strictly positive.)

Although these properties are nice from the point of view of robustness, it was later pointed out (Blackwell 1973) that the Dirichlet process is discrete with probability one.

The most interesting property of the Dirichlet process is the fact that the posterior distribution is very easily described. Indeed, if we define the counting process associated to the observations by

$$N_t^n = \sum_{1 \le k \le n} 1_{\{X_k \le t\}} \tag{3.7}$$

i.e., the number of observations smaller than or equal to t, we have the following result (Ferguson 1973)

PROPOSITION 3.1.

If $F \sim Di(\alpha)$ *and* $X = \{X_k : 1 \leq k \leq n\}$ *satisfies* (2.5) *and* (2.6), *then* $F|X \sim Di(\alpha + N^n)$, *i.e. a posteriori (conditionnally on the observations)* F *is a Dirichlet process with parameter* $\alpha + N^n$.

If $\hat{F}_n(t)$ *denotes the empirical distribution function of the observations, i.e.* $\hat{F}_n(t) = \frac{1}{n} N^n_t$, *as a corollary, we have*

$$E(F_t \mid X) = \frac{\alpha(\infty)}{\alpha(\infty) + n} E(F_t) + \frac{n}{\alpha(\infty) + n} \hat{F}_n(t) \tag{3.8}$$

This result is analytically very nice, since the posterior expectation is a linear function of the prior expectation and of an unbiased estimator. It is in fact a convex combination as it must be (see M. Mouchart and L. Simar 1981).

The question is now : what are the independence properties that explain the analytical flexibility of the Dirichlet process in computing the posterior distribution.

From this point of view, we have a very well known characterization of the Dirichlet process (Wilks 1962, Ferguson 1973).

PROPOSITION 3.2.

$F \sim Di(\alpha)$ *if and only if, for each* n *and for each choice of* $-\infty = t_0 < t_1 < t_2 \ldots < t_n < t_{n+1} = +\infty$, *then*

(i) $$\underset{1 \leq k \leq n}{\perp\!\!\!\perp} \frac{F_{t_k} - F_{t_{k-1}}}{1 - F_{t_{k-1}}}$$

(ii) $$\frac{F_{t_k} - F_{t_{k-1}}}{1 - F_{t_{k-1}}} \sim B[\alpha(t_k) - \alpha(t_{k-1}); \alpha(\infty) - \alpha(t_k)]$$

Instead of extending, as usually done, the characterizations of Dirichlet distribution to the Dirichlet process, we will derive them directly.

To this aim, as in the general framework of stochastic processes, we define the following σ-algebras, for $-\infty \leq s < t \leq +\infty$

$$G^t_s = \sigma\left\{\frac{F_u - F_s}{F_t - F_s} : s < u < t\right\} \tag{3.9}$$

and in particular,

$$G^t_{-\infty} = \sigma\left\{\frac{F_u}{F_t} : -\infty < u < t\right\} \tag{3.10}$$

$$\mathcal{G}_s^\infty = \sigma\left\{\frac{F_u - F_s}{1 - F_s} : s < u < \infty\right\}$$

Note that $\dfrac{F_u - F_s}{F_t - F_s}$; $s < u < t$, is the distribution of the observation conditionnally on the fact that this observation is in the interval $(s,t]$ (if $F_t - F_s = 0$, $\dfrac{F_u - F_s}{F_t - F_s}$ will be defined as zero).

Therefore, the σ-algebra $\mathcal{G}_s^t$ represents the information contained in the knowledge of the conditional distribution process on the interval $(s,t]$. And so, for instance, if $F_{-\infty}^t$ is the past of the distribution process, then $\mathcal{G}_{-\infty}^t$ is the conditional past of the distribution process.

In writing expressions involving σ-algebras, we will not distinguish the σ-algebra generated by a random variable from this random variable, *e.g.*, we will write F_t instead of F_t^t. For instance let us remark that for each $-\infty \leqslant s < t \leqslant +\infty$

$$F_s^t = \mathcal{G}_s^t \vee F_s \vee F_t \tag{3.11}$$

With these definitions, Proposition 3.2 may be rephrased as follows : the Dirichlet process is a distribution process whose past is independent of the conditional future. More precisely

PROPOSITION 3.3

$F \sim \mathrm{Di}(\alpha)$ *if and only if for each choice of* $-\infty < s < t < +\infty$,

(i) $\dfrac{F_t - F_s}{1 - F_s} \perp\!\!\!\perp F_{-\infty}^s$

and

(ii) $\dfrac{F_t - F_s}{1 - F_s} \sim B[\alpha(t) - \alpha(s), \alpha(\infty) - \alpha(t)]$

Note that (i) *is also equivalent to*

(iii) $F_{-\infty}^s \perp\!\!\!\perp \mathcal{G}_s^\infty$

The proof follows from the trivial fact that

$$\sigma\left\{\frac{F_{t_k} - F_{t_{k-1}}}{1 - F_{t_{k-1}}} : \ell+1 \leqslant k \leqslant m\right\} = \sigma\left\{\frac{F_{t_k} - F_{t_\ell}}{1 - F_{t_\ell}} : \ell+1 \leqslant k \leqslant m\right\}$$

We can go one step further using the following lemma whose proof consists of a simple change of variables (see Wilks 1962).

LEMMA 3.4.

If $U_1 \perp\!\!\!\perp U_2$, $U_1 \sim B(a,b)$, $U_2 \sim B(a+b,c)$, *if we set* $U_1' = \dfrac{1-U_2}{1-U_1U_2}$ $U_2' = 1 - U_1U_2$, *then* $U_1' \perp\!\!\!\perp U_2'$ $U_1' \sim B(c,b)$, $U_2' \sim B(b+c,a)$.

This lemma will be frequently used in the sequel. Firstly, it allows us to see that the Dirichlet process is a distribution process whose conditional reduction on any interval is independent of the past and the future. More precisely

PROPOSITION 3.5.

$F \sim Di(\alpha)$ *if and only if for each choice of* $-\infty \leqslant s < u < t \leqslant +\infty$,

$$\text{(i)} \qquad \frac{F_u - F_s}{F_t - F_s} \perp\!\!\!\perp (F_{-\infty}^{s} \vee F_t^{+\infty})$$

and

$$\text{(ii)} \qquad \frac{F_u - F_s}{F_t - F_s} \sim B[\alpha(u) - \alpha(s); \alpha(t) - \alpha(u)].$$

Note that (i) *is also equivalent to*

$$\text{(iii)} \qquad G_s^t \perp\!\!\!\perp (F_{-\infty}^{s} \vee F_t^{\infty}).$$

Proof :

(i) and (ii) are clearly sufficient since it suffices to take $t = \infty$ to obtain Proposition 3.3. From Proposition 3.3 we know that $\dfrac{F_t - F_u}{1 - F_u} \perp\!\!\!\perp F_{-\infty}^{u}$ $\perp\!\!\!\perp G_t^{\infty}$ and that $\dfrac{F_u - F_s}{1 - F_s} \perp\!\!\!\perp F_{-\infty}^{s}$. Hence $\dfrac{F_u - F_s}{1 - F_s} \perp\!\!\!\perp \dfrac{F_t - F_u}{1 - F_u} \perp\!\!\!\perp F_{-\infty}^{s} \perp\!\!\!\perp G_t^{\infty}$. But $1 - \dfrac{F_u - F_s}{1 - F_s} = \dfrac{1 - F_u}{1 - F_s} \sim B[\alpha(\infty) - \alpha(u), \alpha(u) - \alpha(s)]$ and $1 - \dfrac{F_t - F_u}{1 - F_u} = \dfrac{1 - F_t}{1 - F_u} \sim B[\alpha(\infty) - \alpha(u), \alpha(t) - \alpha(u)]$. Applying Lemma 3.4 with $U_1 = \dfrac{1 - F_t}{1 - F_u}$ and $U_2 = \dfrac{1 - F_u}{1 - F_s}$, we see that $U_2' = 1 - \dfrac{1 - F_t}{1 - F_u}\dfrac{1 - F_u}{1 - F_s} = \dfrac{F_t - F_s}{1 - F_s}$ is independent of $U_1' = \dfrac{F_u - F_s}{1 - F_s}\dfrac{1 - F_s}{F_t - F_s} = \dfrac{F_u - F_s}{F_t - F_s}$, $\dfrac{F_t - F_s}{1 - F_s} \sim B[\alpha(t) - \alpha(s),$ $\alpha(\infty) - \alpha(t)]$ and $\dfrac{F_u - F_s}{F_t - F_s} \sim B[\alpha(u) - \alpha(s), \alpha(t) - \alpha(u)]$. This is (ii). Moreover $\dfrac{F_u - F_s}{F_t - F_s} \perp\!\!\!\perp \dfrac{F_t - F_s}{1 - F_s} \perp\!\!\!\perp F_{-\infty}^{s} \perp\!\!\!\perp G_t^{\infty}$ and this implies (i) since $\dfrac{F_t - F_s}{1 - F_s} \vee F_{-\infty}^{s} \vee G_t^{\infty} = F_{-\infty}^{s} \vee F_t^{\infty}$. From (i) we easily obtain by induction that for each choice of $s < u_1 < u_2 \dots < u_n < t$, $u_0 = s$, then

$\underset{1\leq k\leq n}{\perp\!\!\!\perp} \frac{F_{u_k} - F_{u_{k-1}}}{F_t - F_{u_{k-1}}} \perp\!\!\!\perp (F_{-\infty}^{s} \vee F_t^{\infty})$ and since $\sigma\left\{\frac{F_{u_k} - F_{u_{k-1}}}{F_t - F_{u_{k-1}}} : 1 \leq k \leq n\right\} = \sigma\left\{\frac{F_{u_k} - F_s}{F_t - F_s} : 1 \leq k \leq n\right\}$ we see that (iii) is a direct consequence of (i).

□

By taking $s = -\infty$, we get as a corollary another property of a Dirichlet process that is also sufficient for characterizing it.

PROPOSITION 3.6.

$F \sim Di(\alpha)$ *if and only if for each choice of* $-\infty < s < t < +\infty$,

(i) $\quad \frac{F_s}{F_t} \perp\!\!\!\perp F_t^{\infty}$

and

(ii) $\quad \frac{F_s}{F_t} \sim B[\alpha(s);\ \alpha(t) - \alpha(s)]$.

Note that (i) *is also equivalent to*

(iii) $\quad G_{-\infty}^{t} \perp\!\!\!\perp F_t^{\infty}$.

Hence the Dirichlet process is a distribution process whose conditional past is independent of the future.

Proof

Clearly (i) implies that for each choice of $-\infty = t_0 < t_1 < t_2 \ldots < t_n < t_{n+1} = \infty$, $\underset{1\leq k\leq n}{\perp\!\!\!\perp} \frac{F_{t_k}}{F_{t_{k+1}}}$. Hence (iii) is a direct consequence of (i) since $\sigma\left\{\frac{F_{t_{k-1}}}{F_{t_k}} : \ell+1 \leq k \leq m\right\} = \sigma\left\{\frac{F_{t_k}}{F_{t_m}} : \ell \leq k \leq m-1\right\}$. Now by (iii) it follows that $\forall -\infty \leq s < u < t \leq +\infty$, $G_{-\infty}^{s} \perp\!\!\!\perp \frac{F_s}{F_u} \perp\!\!\!\perp \frac{F_u}{F_t} \perp\!\!\!\perp F_t^{\infty}$, $\frac{F_s}{F_u} \sim B[\alpha(s), \alpha(u) - \alpha(s)]$ and $\frac{F_u}{F_t} \sim B[\alpha(u), \alpha(t) - \alpha(u)]$. Applying Lemma 3.4, with $U_1 = \frac{F_s}{F_u}$ and $U_2 = \frac{F_u}{F_t}$, we have $1 - U_2' = \frac{F_s}{F_t}$ is independent of $1 - U_1' = \frac{F_u - F_s}{F_t - F_s}$, $\frac{F_s}{F_t} \sim B[\alpha(s), \alpha(t) - \alpha(s)]$ and $\frac{F_u - F_s}{F_t - F_s} \sim B[\alpha(u) - \alpha(u), \alpha(t) - \alpha(u)]$ (3.4(ii)). Hence $G_{-\infty}^{s} \perp\!\!\!\perp \frac{F_s}{F_t} \perp\!\!\!\perp \frac{F_u - F_s}{F_t - F_s} \perp\!\!\!\perp F_t^{\infty}$ and since $G_{-\infty}^{s} \vee \frac{F_s}{F_t} \vee F_t^{\infty} \vee F_{-\infty}^{s} \vee F_t^{\infty}$ we obtain 3.4 (i).

□

As a consequence of these results, one may define distribution processes that generalize Dirichlet processes. These processes, called *neutral*, will be the object of the next section.

4. NEUTRAL PROCESSES

Using Propositions 3.3, 3.5 and 3.6, we may generalize the Dirichlet process to distribution processes that are defined only through independence relations.

DEFINITION 4.1.

A distribution process F *is called*

(i) *neutral to the right if* $\forall -\infty < s < t < +\infty$

$$\frac{F_t - F_s}{1 - F_s} \perp\!\!\!\perp F^s_{-\infty}$$

(ii) *neutral to the left if* $\forall -\infty < s < t < +\infty$

$$\frac{F_s}{F_t} \perp\!\!\!\perp F^{\infty}_t$$

(iii) *neutral if* $\forall -\infty \leq s < u < t \leq +\infty$

$$\frac{F_u - F_s}{F_t - F_s} \perp\!\!\!\perp (F^s_{-\infty} \vee F^{\infty}_t)$$

As a direct consequence of the parts (iii) of those propositions, we have the following result

PROPOSITION 4.2.

A distribution process F *is*

(i) *neutral to the right, if and only if,* $\forall -\infty < t < +\infty$

$$F^t_{-\infty} \perp\!\!\!\perp G^{\infty}_t$$

(ii) *neutral to the left, if and only if,* $\forall -\infty < t < +\infty$

$$G^t_{-\infty} \perp\!\!\!\perp F^{\infty}_t$$

(iii) *neutral, if and only if,* $\forall -\infty \leq s < t \leq +\infty$

$$G^t_s \perp\!\!\!\perp (F^s_{-\infty} \vee F^{\infty}_t)$$

These definitions may then be interpreted as follows : a neutral to the right process is a distribution process whose past is independent of the conditional future while a neutral to the left process is a distribution process whose future is independent of the conditional past. A neutral process is a distribution process whose conditional reduction to any (closed) interval is independent of the past and the future.

Equivalent definitions of neutral to the right and neutral to the left processes were first given by Docksum (1974). The definition of a neutral process is new but we will see that it characterizes completely the Dirichlet process.

Clearly a neutral process is neutral to the right ($t = \infty$) and neutral to the left ($s = -\infty$) but is it true that a process neutral to the right and neutral to the left is neutral ? Docksum (1974) conjectured that such a process is necessarily a Dirichlet process. This would imply that neutral processes are Dirichlet processes. But we are able to prove that there is a simpler way than Docksum's conjecture to characterize Dirichlet processes only by independence relations. Indeed, a neutral to the right and neutral to the left process has clearly the following properties :

$\forall\ -\infty < s < t < +\infty$

$$\frac{F_t - F_s}{1 - F_s} \perp\!\!\!\perp F_s \tag{4.1}$$

and

$$\frac{F_s}{F_t} \perp\!\!\!\perp F_t \tag{4.2}$$

or equivalently,

$\forall\ -\infty < s < t < +\infty$

$$\frac{1 - F_t}{1 - F_s} \perp\!\!\!\perp 1 - F_s \tag{4.3}$$

and

$$\frac{F_s}{F_t} \perp\!\!\!\perp F_t \tag{4.4}$$

In the proofs of Propositions 3.5 and 3.6, we only used Lemma 3.4 to prove that one of these independence was implying the other. The key fact is that Lemma 3.4 has some kind of converse.

LEMMA 4.3.

Let U_1 *and* U_2 *be non degenerate random variables with values in* $[0,1]$ *and let* $U_1' = \frac{1 - U_2}{1 - U_1U_2}$, $U_2' = 1 - U_1U_2$ (U_1' *is defined to be zero if* $U_1 = U_2 = 1$), *then* $U_1 \perp\!\!\!\perp U_2$ *and* $U_1' \perp\!\!\!\perp U_2'$ *imply that there exist* a,b,c *positive real numbers, such that* $U_1 \sim B(a,b)$, $U_2 \sim B(a+b,c)$, $U_1' \sim B(c,b)$ *and* $U_2' \sim B(b+c,a)$.

The proof of this lemma is quite similar to the proof given by Fabius to characterize Dirichlet distribution (Fabius 1973) and, in return, this lemma suffices to prove the two characterizations of Dirichlet distribution given in this article and the characterization given by James and Mosimann (1980).

Unlike Docksum's article (1974) and Ferguson's article (1974), we only meet two kinds of degenerate distribution processes in our framework.

Following Ferguson's article, the three trivial types of distribution processes are

T_1 : F is non random, *i.e.* $F_t = \mu(t)$ $\forall -\infty < t < +\infty$ *a.s.*

T_2 : F is degenerate at a random point, *i.e.* $F_t = 1_{\{Y \le t\}}$ $\forall -\infty < t < +\infty$ *a.s.* and Y has distribution $\mu(t)$

T_3 : F is degenerate at two non random points, *i.e.* $F_t = U\, 1_{[t_0,t_1)}(t) + 1_{[t_1,\infty]}(t)$ $\forall -\infty < t < +\infty$ *a.s.* where $t_0 < t_1$ and U is an arbitrary random variable with values in $[0,1]$.

Using Lemma 4.3, we obtain the following result which proves, as a corollary, Docksum's conjecture.

PROPOSITION 4.4.

If a distribution process F *is neutral to the right and satisfies* (4.4) $\forall -\infty < s < t < +\infty$ *or is neutral to the left and satisfies* (4.3) $\forall -\infty < s < t < +\infty$, *then* F *is a Dirichlet process of types* T_1 *or* T_3.

The neutral processes retain some of the properties of the Dirichlet process. In particular, we have good analytical insights into the posterior distribution, as shown in the following proposition that is due to Docksum (1974). We will however give another proof to show the advantage of our framework.

PROPOSITION 4.5.

Let F *be a distribution process and* $X = \{X_k : 1 \leqslant k \leqslant n\}$ *a random vector, satisfying* (2.5) *and* (2.6). *If* F *is neutral (resp. neutral to the right, resp. neutral to the left), then conditional on* X, F *is still neutral (resp. neutral to the right, resp. neutral to the left).*

Proof

Since (2.5) implies that the distribution of X_{k+1} conditional on $(F_{-\infty}^{+\infty} \vee X_1^k)$ is the same as the distribution of X_{k+1} conditional on $F_{-\infty}^{+\infty}$

by induction, it suffices to prove the result for $n = 1$. Now,

$$E[f(X_1) \mid F_{-\infty}^{+\infty}] = \int_{-\infty}^{+\infty} f(\tau)\, dF_\tau$$

$$= \int_{-\infty}^{s} f(\tau)\, dF_\tau + (F_t - F_s) \int_s^t f(\tau)\, d\,\frac{F_\tau - F_s}{F_t - F_s}$$

$$+ \int_t^\infty f(\tau)\, dF_\tau .$$

So if $Y \in G_s^t$ and $Z \in F_{-\infty}^{s} \vee F_t^{\infty}$, using 4.3,

$$E[YZ\, f(X_1)] = E(Y)\, E\left[Z \left\{\int_{-\infty}^{s} f(\tau)\, dF_\tau + \int_t^\infty f(\tau)\, dF_\tau\right\}\right]$$

$$+ E\left[Y \int_s^t f(\tau)\, d\,\frac{F_\tau - F_s}{F_t - F_s}\right] E[Z(F_t - F_s)]$$

$$= E(Y)\, E[Z\, f(X_1);\ X_1 \notin (s,t]]$$

$$+ \frac{E[Z(F_t - F_s)]}{E[F_t - F_s]}\, E[Yf(X_1);\ X_1 \in (s,t]]$$

This implies that

$$E(YZ \mid X_1) = E(Y)\, E(Z \mid X_1) \qquad \text{on } \{s < X_1 \leq t\}^c$$

$$E(Y \mid X_1)\, \frac{E[Z(F_t - F_s)]}{E(F_t - F_s)} \qquad \text{on } \{s < X_1 \leq t\}$$

So $\qquad G_s^t \perp\!\!\!\perp (F_{-\infty}^{s} \vee F_t^{\infty}) \mid X_1$

□

Note that the proof shows how a neutral process is revised by an observation. Indeed since $E(Y \mid X_1) = E(Y)$ on $\{s < X_1 \leq t\}^c$, G_s^t is not revised by an observation falling outside $(s,t]$, and since

$$E(Z \mid X_1) = \frac{E[Z(F_t - F_s)]}{E[(F_t - F_s)]} = E(Z \mid s < X_1 \leq t)$$

on $\{s < X_1 \leq t\}$, the past and the future of the distribution process, *i.e.* $F_{-\infty}^{s} \vee F_t^{\infty}$, is revised by an observation falling in the interval $(s,t]$, only by the fact that this observation is in the interval and not by the exact value of this observation.

As a direct consequence of Proposition 4.4, the only thing needed to characterize completely the posterior distribution of a process neutral to the right (resp. neutral to the left) is to find the posterior distribution of $\frac{F_t - F_s}{1 - F_s}$ (resp. $\frac{F_s}{F_t}$) $(\forall\ -\infty < s < t < \infty)$ for this distribution entirely specifies the distribution of such a process.

Since neutral processes are essentially Dirichlet processes, it is interesting to give a proof of Ferguson's theorem (1973) (Proposition 3.1) by only using independence properties of neutral processes and to reconstruct, in that way, the Beta distribution. This gives the following proposition.

PROPOSITION 4.6.

Let F *be a neutral process and* $X = \{X_k : 1 \leq k \leq n\}$ *a random vector satisfying* (2.5) *and* (2.6). *If* $\zeta_{s,t}$ *denotes the distribution of the random variable* $\frac{F_t - F_s}{1 - F_s}$, *then*

$$\zeta_{s,t}(dy \mid X) = \frac{y^{N_t^n - N_s^n}(1-y)^{n - N_t^n}}{\mu_{s,t}[N_t^n - N_s^n, n - N_t^n]} \zeta_{s,t}(dy)$$

where

$$\mu_{s,t}(a,b) = E\left[\left(\frac{F_t - F_s}{1 - F_s}\right)^a \left(1 - \frac{F_t - F_s}{1 - F_s}\right)^b\right]$$

Proof :

Let $Y = \frac{F_t - F_s}{1 - F_s}$ and $X_1^k = \sigma\{X_\ell : 1 \leq \ell \leq k\}$.

Since $E[g(X_{k+1}) \mid F_{-\infty}^{+\infty} \vee X_1^k] = E[g(X_{k+1} \mid F_{-\infty}^{+\infty}]$

$$= \int_{-\infty}^{+\infty} g(\tau)\, dF_\tau$$

$$= \int_{-\infty}^{s} g(\tau)\, dF_\tau + (F_t - F_s)\int_s^t g(\tau)\, d\,\frac{F_\tau - F_s}{F_t - F_s} + (1 - F_t)\int_t^\infty g(\tau)\, d\,\frac{F_\tau - F_t}{1 - F_t}$$

using Propositions 4.2 and 4.5, we get

$$E[f(Y)\, g(X_{k+1}) \mid X_1^k] = E[f(Y) \mid X_1^k]\, E\left[\int_{-\infty}^{s} g(\tau)\, dF_\tau \mid X_1^k\right]$$

$$+ E[f(Y)\{F_t - F_s\} \mid X_1^k]\, E\left[\int_s^t g(\tau)\, d\,\frac{F_\tau - F_s}{F_t - F_s} \mid X_1^k\right]$$

$$+ E[f(Y)\{1 - F_t\} \mid X_1^k]\, E\left[\int_t^\infty g(\tau)\, d\,\frac{F_\tau - F_t}{1 - F_t} \mid X_1^k\right]$$

$$= E[f(Y) \mid X_1^k]\, E[g(X_{k+1})\, 1_{\{X_{k+1} \leq s\}} \mid X_1^k]$$

$$+ \frac{E[f(Y)\{F_t - F_s\} \mid X_1^k]}{E[F_t - F_s \mid X_1^k]}\, E\left[g(X_{k+1})\, 1_{\{s < X_{k+1} \leq t\}} \mid X_1^k\right]$$

$$+ \frac{E[f(Y)\{1 - F_t\} \mid X_1^k]}{E[1 - F_t \mid X_1^k]}\, E\left[g(X_{k+1})\, 1_{\{X_{k+1} > t\}} \mid X_1^k\right]$$

Since,

$$E[f(Y)\ \{F_t - F_s\} \mid X_1^k] = E[f(Y)\ Y \mid X_1^k]\ E[1 - F_s \mid X_1^k]$$

and

$$E[f(Y)\ \{1 - F_t\} \mid X_1^k] = E[f(Y)\ (1 - Y) \mid X_1^k]\ E[1 - F_s \mid X_1^k]$$

we get,

$$E[f(Y) \mid X_1^{k+1}] = \begin{cases} E[f(Y) \mid X_1^k] & \text{on } \{X_{k+1} \leq s\} \\ \dfrac{E[f(Y)\ Y \mid X_1^k]}{E(Y \mid X_1^k)} & \text{on } \{s < X_{k+1} \leq t\} \\ \dfrac{E[f(Y)\ \{1 - Y\} \mid X_1^k]}{E(1 - Y \mid X_1^k)} & \text{on } \{X_{k+1} > t\} \end{cases}$$

Now if the proposition is true for k,

$$E[f(Y) \mid X_1^k] = \frac{E[f(Y)\ Y^a\ (1-Y)^b]}{E[Y^a\ (1-Y)^b]} \qquad \text{on } \{N_t^k - N_s^k = a,\ k - N_t^k = b\}$$

it is easy to see that this formula is still true for $k+1$, since if $X_{k+1} \leq s$, $N_s^{k+1} = N_s^k + 1$ and $N_t^{k+1} = N_t^k + 1$, if $s < X_{k+1} \leq t$, $N_s^{k+1} = N_s^k$ and $N_t^{k+1} = N_t^k + 1$, and if $X_{k+1} > t$, $N_s^{k+1} = N_s^k$ and $N_t^{k+1} = N_t^k$.

Therefore, if F is a Dirichlet process with parameter α, Proposition 4.6 implies that, conditional on X, F is a Dirichlet process with parameter $\alpha + N^n$.

It remains to show that we also have tractable analytical insights into the posterior distribution of neutral to the right or neutral to the left processes. First of all, let us remark that it suffices to consider neutral to the right processes, since F is neutral to the left, if and only if $\tilde{F}$ is neutral to the right, where

$$\tilde{F}_t = 1 - F_{(-t)-}.$$

Furthermore, to describe this posterior distribution, it is usefull to recall the following characterization of neutral to the right processes in terms of increasing processes with independent increments.

PROPOSITION 4.7. (Docksum 1974)

If F *is a neutral to the right process and*

$$A_t = \ell n \frac{1}{1 - F_t} \tag{4.5}$$

then A *is a positive increasing right continuous process with independent increments and such that* $\lim_{t \to -\infty} A_t = 0$ *and* $\lim_{t \to +\infty} A_t = +\infty$. *Conversely, if* A *is such a process and*

$$F_t = 1 - e^{-A_t} \tag{4.6}$$

then F *is a neutral to the right process.*

Now, if we define the following σ-algebras

$$B_s^t = \sigma\{A_u - A_s : s < u \leq t\} \tag{4.7}$$

the independent increments property may be written as

$$A_t - A_s \perp\!\!\!\perp B_{-\infty}^s \quad \forall -\infty < s < t < +\infty \tag{4.8}$$

and this is equivalent to

$$B_{-\infty}^s \perp\!\!\!\perp B_s^{\infty} \qquad \forall -\infty < s < +\infty \tag{4.9}$$

Clearly, (4.8) is a rewriting of the definition of a neutral to the right process. The advantage comes from the fact that, as will be seen later, increasing processes with independent increments have been extensively studied. The jumps of these processes are known to play a crucial role. Therefore it is important to introduce these jumps in the independence relations.

If we define the strict past of the process, *i.e.* the σ-algebra of events happening strictly before t by

$$B_{-\infty}^{t-} = \sigma\{A_s : -\infty < s < t\} \tag{4.10}$$

then

$$B_{-\infty}^t = B_{-\infty}^{t-} \vee (A_t - A_{t-}) \tag{4.11}$$

and

$$B_{-\infty}^{t-} \perp\!\!\!\perp (A_t - A_{t-}) \tag{4.12}$$

Combining (4.12) with (4.9), we obtain $\forall -\infty < t < +\infty$

$$B_{-\infty}^{t-} \perp\!\!\!\perp (A_t - A_{t-}) \perp\!\!\!\perp B_t^{\infty} \tag{4.13}$$

By virtue of Proposition 4.5, it follows that if X satisfies (2.6), we also have, $\forall -\infty < t < +\infty$

$$B_{-\infty}^{t-} \perp\!\!\!\perp (A_t - A_{t-}) \perp\!\!\!\perp B_t^{\infty} \mid X \tag{4.14}$$

What is more remarkable is that (4.14) is still true if t is replaced by X. Indeed, if we define, the σ-algebra of events happening strictly before X by

$$\begin{aligned} B_{-\infty}^{X-} &= \bigvee_{-\infty \le t < +\infty} \sigma\{B_{-\infty}^{t} \cap \{X > t\}\} \\ &= \sigma\Big\{\bigcup_{-\infty < t < +\infty} B_{-\infty}^{t} \cap \{X > t\}\Big\} \end{aligned} \tag{4.15}$$

and the σ-algebra of events posterior to X by

$$\begin{aligned} B_X^{\infty} &= \sigma\{A_{X+u} - A_X : 0 < u < \infty\} \vee \sigma(X) \\ &= \sigma\{(A_t - A_X)\, 1_{\{X<t\}} : -\infty < t < +\infty\} \vee \sigma(X) \end{aligned} \tag{4.16}$$

then we have the following proposition.

<u>*PROPOSITION 4.8.*</u>

If A *is a positive increasing right continuous process with independent increments with* $A_{-\infty} = 0$ *and* $A_{+\infty} = \infty$ *and if* X *is a real random variable such that* $P(X \leq t \mid B_{-\infty}^{+\infty}) = 1 - e^{-A_t}$, *then*

$$B_{-\infty}^{X-} \perp\!\!\!\perp (A_X - A_{X-}) \perp\!\!\!\perp B_X^{\infty} \mid X \tag{4.17}$$

<u>Proof</u> :

Since $\bigcup_t B_{-\infty}^{t} \cap \{X > t\}$ is a π-system it suffices to prove 4.17 for $Y_t\, 1_{\{X>t\}}$ where $Y_t \in B_{-\infty}^{t}$, $e^{-u(A_X - A_{X-})}$, and $Z_X = \prod_{1\le i\le n} e^{-v_i(A_{X+u_i} - A_X)}$.

Now,

$$= E[g(X)\, Y_t\, 1_{\{X>t\}}\, e^{-u(A_X - A_{X-})}\, Z_X]$$

$$= E\Big[Y_t \int_t^{\infty} g(\tau)\, e^{-u(A_\tau - A_{\tau-})}\, Z_\tau\, d(1 - e^{-A_\tau})\Big]$$

$$= E\Big[Y_t\, e^{-A_t}\Big]\, E\Big[\int_t^{\infty} g(\tau)\, e^{-u(A_\tau - A_{\tau-})}\, Z_\tau\, d\Big(1 - e^{-(A_\tau - A_t)}\Big)\Big]$$

since $d\left(1 - e^{-A_\tau}\right) = e^{-A_t} d\left(1 - e^{-(A_\tau - A_t)}\right)$ for $\tau > t$;

$$= \frac{E[Y_t\, e^{-A_t}]}{E[e^{-A_t}]} E\left[\int_t^\infty g(\tau)\, e^{-u(A_\tau - A_{\tau-})} E(Z_\tau)\, d(1 - e^{-A_\tau})\right]$$

since $Z_\tau \in \mathcal{B}_\tau^\infty$

$$= \frac{E[Y_t\, e^{-A_t}]}{E[e^{-A_t}]} E\left[g(X)\, e^{-u(A_X - A_{X-})} h(X);\, X > t\right]$$

where $h(\tau) = E(Z_\tau)$. Therefore

$$E\left[Y_t\, 1_{\{X>t\}}\, e^{-u(A_X - A_{X-})} Z_X \mid X\right] \tag{4.18}$$

$$= \frac{E[Y_t\, e^{-A_t}]}{E(e^{-A_t})} 1_{\{X>t\}} h(X)\, E\left[e^{-u(A_X - A_{X-})} \mid X\right]$$

which yields the result.

□

The last equality contains more information than Proposition 4.8 and will be used in the next section.

5. THE POSTERIOR DISTRIBUTION OF A NEUTRAL TO THE RIGHT PROCESS

The computation of the posterior distribution of a neutral to the right process has been first partially made by Docksum (1974). Later Ferguson (1974) gave an alternative description of Docksum's results in terms of the distribution of the process A. In this section, we will give an alternative computation of the Laplace transform of the posterior distribution of A_t. In the light of Propositions 4.5 and 4.7, this is clearly sufficient to characterize the posterior distribution of a neutral to the right process. The use of Proposition (4.8) will allow us to reduce this computation to the posterior distribution of the jumps at the observations. By induction, it will suffice to compute the Laplace transform of the posterior distribution resulting from a single observation X satisfying

$$E\left[f(X) \mid \mathcal{B}_{-\infty}^{+\infty}\right] = \int_{-\infty}^{+\infty} f(\tau)\, d(1 - e^{-A_\tau}). \tag{5.1}$$

This computation depends crucially on the following fundamental increasing right continuous function which specifies the prior distribution,

$$\psi(u,t) = -\ell n\, E(e^{-uA_t}). \tag{5.2}$$

Note that

$$\psi(u,t-) = -\ell n\, E(e^{-uA_{t-}}). \tag{5.3}$$

The problem consists in computing $E(e^{-uA_t} \mid X)$. We have already obtained some results along these lines in Proposition 4.8 and in its proof.

Indeed, on $\{t < X\}$

$$E\left[e^{-uA_t} \mid X\right] = \frac{E[e^{-uA_t}\, e^{-A_t}]}{E[e^{-A_t}]} \tag{5.4}$$

and so

$$E\left[e^{-uA_t} \mid X\right] = e^{-\psi(u+1,t) + \psi(1,t)}. \tag{5.5}$$

On $\{t > X\}$

$$E\left[e^{-uA_t} \mid X\right] = E\left[e^{-u(A_t - A_X)} \mid X\right] E\left[e^{-u(A_X - A_{X-})} \mid X\right] E\left[e^{-uA_{X-}} \mid X\right]. \tag{5.6}$$

But since $(A_t - A_X)\, 1_{\{t>X\}} \in \mathcal{B}_X^{\infty}$, it results from the proof of Proposition 4.8, that

$$E\left[e^{-u(A_t - A_X)} \mid X\right] = e^{-\psi(u,t) + \psi(u,X)}. \tag{5.7}$$

On the other hand,

$$E\left[e^{-uA_{X-}} \mid X\right] = e^{-\psi(u+1,X-) + \psi(1,X-)}. \tag{5.8}$$

Note that (5.6) relies on the fact that $A_{X-} \in \mathcal{B}_{-\infty}^{X-}$ and that (5.8) is, in same sense, the limit of (5.5) as t increases to X strictly from below. These two facts may be verified as follows. Given $-\infty < s < t < +\infty$, if $\tau_{n,k} = s + \frac{k}{2^n}(t-s)$, let us define

$$A^1_{n,\tau} = \sum_{1 \le k \le 2^n} A_{\tau_{n,k-1}}\, 1_{\{\tau_{n,k-1} < \tau \le \tau_{n,k}\}}$$

and

$$A^2_{n,\tau} = \sum_{1 \le k \le 2^n} A_{\tau_{n,k}}\, 1_{\{\tau_{n,k-1} < \tau \le \tau_{n,k}\}}.$$

It is easy to see that, on $(s,t]$, $A^1_{n,\tau}$ increases strictly from below to $A_{\tau-}$ and $A^2_{n,\tau}$ decreases to A_τ as n tends to infinity. And so if $s < X \leq t$, $A_{X-} = \lim_{n\to\infty} A^1_{n,X}$ which proves the first fact and using (5.5) we easily obtain

$$E[e^{-uA_{X-}} \mid X] =$$

$$= \lim_{n\to\infty} \sum_{1\le k\le 2^n} E\left[e^{-uA_{\tau_{n,k-1}}} \mid X\right] 1_{\{\tau_{n,k-1}<X\le\tau_{n,k}\}}$$

$$= \lim_{n\to\infty} \sum_{1\le k\le 2^n} e^{-\psi(u+1,\tau_{n,k-1})+\psi(1,\tau_{n,k-1})} 1_{\{\tau_{n,k-1}<X\le\tau_{n,k}\}}$$

$$= e^{-\psi(u+1,X-)+\psi(1,X-)}$$

The problem of computing the posterior distribution is then reduced to the computation of the Laplace transform of the posterior distribution of the jump of the process A at the observation X, *i.e.*

$$E\left[e^{-u(A_X-A_{X-})} \mid X\right].$$

Let us first remark that the function $\psi(u,t)$ may be extended to a σ-finite measure on $(R,\mathcal{B})$ for each $u \in (0,\infty)$. The key fact of the computation is that all these measures are equivalent and the result of the computation will be expressed in terms of the Radon-Nikodym derivative of $\psi(u,t)$ with respect to $\psi(1,t)$.

The proof of this requires the representation of Levy and Itô of a positive increasing right continuous processes with independent increments (see for instance Breiman, 1968). We will sketch it briefly.

First let

$$J = \{t \in R : \psi(1,t-) < \psi(1,t)\}. \tag{5.9}$$

This set is countable and we remark that

$$\begin{aligned} J &= \{t \in R : \mu(t-) < \mu(t)\} \\ &= \{t \in R : \psi(u,t-) < \psi(u,t)\}. \end{aligned} \tag{5.10}$$

We may then decompose $\psi(u,t)$ into its discrete and continuous parts as follows

$$\psi(u,t) = \sum_{\substack{s\le t\\ s\in J}} \Delta\psi(u,s) + \psi_1(u,t) \tag{5.11}$$

where

$$\Delta\psi(u,s) = \psi(u,s) - \psi(u,s-). \tag{5.12}$$

Next if we define, $\forall\ -\infty < t < +\infty$, $\forall\ B \in \mathcal{B}^+$ bounded away from zero, ($\mathcal{B}^+$ is the σ-algebra of borel sets on $R^+ = (0,\infty)$).

$$\nu_t(B) = E \sum_{\substack{s\le t \\ s\in J^c}} 1_B(A_s - A_{s-}) \tag{5.13}$$

then for each t, $\nu_t(\cdot)$ is a σ-finite measure on $(R^+,\mathcal{B}^+)$ and for each $B \in \mathcal{B}^+$ bounded away from zero, $\nu_\cdot(B)$ is a continuous increasing function on R.

We have then the following representation of $\psi_1(u,t)$

$$\psi_1(u,t) = ua(t) + \int_0^\infty (1 - e^{-ux})\, \nu_t\,(dx) \tag{5.14}$$

where $a(t)$ is a positive increasing continuous function on R;

Note also that for each $t \in R$,

$$\int_0^\infty (x \wedge 1)\, \nu_t\,(dx) < \infty \quad \text{and} \tag{5.15}$$

$$a(t) = \lim_{u\to\infty} \frac{1}{u}\psi_1(u,t). \tag{5.16}$$

Moreover

$$\lim_{t\to-\infty} \int_0^\infty (x \wedge 1)\, \nu_t\,(dx) = 0 \tag{5.17}$$

$$\lim_{t\to-\infty} a(t) = 0. \tag{5.18}$$

Now, since $\psi_1(u,t)$ is a continuous increasing function in t, it can be extended to a measure on $(R,\mathcal{B})$. The same is true for $a(t)$. Similarly if we define, $\forall\, -\infty < s < t < +\infty$ $\forall\, B \in \mathcal{B}^+$ bounded away from zero

$$\nu[(s,t] \times B] = \nu_t(B) - \nu_s(B) \tag{5.19}$$

then ν can be extended to a measure on $(R \times R^+,\ \mathcal{B} \otimes \mathcal{B}^+)$ and we have the following formula for all positive Borel functions f

$$\int_R f(t)\, d\,\psi_1(u,t) = u\int_R f(t)\, da(t) + \int_{R\times R^+} (1-e^{-ux})\, f(t)\, \nu(dt,dx) \tag{5.20}$$

Hence, it is clear that for $u > 0$, $\psi_1(u,\cdot)$ is absolutely continuous with respect to $\psi_1(1,\cdot)$ and vice versa. So for each $u > 0$, there exists a positive borel function $\varphi(u,t)$ such that

$$\psi_1(u,t) = \int_{-\infty}^t \varphi(u,s)\, d\psi_1(1,s) \tag{5.21}$$

We are now ready to compute the jump of the posterior distribution at the observation.

PROPOSITION 5.1.

Let A *be a positive increasing right continuous process with independent increments, with* $A_{-\infty} = 0$ *and* $A_{\infty} = \infty$ *and let* X *be a real random variable such that*

$$P(X \leq t \mid \mathcal{B}_{-\infty}^{+\infty}) = 1 - e^{-A_t} \qquad \forall\, t \in R$$

then

$$E\left[e^{-u(A_X - A_{X-})} \mid X\right] = h(u+1,X) - h(u,X)$$

where

$$h(u,t) = \varphi(u,t)\, 1_{J^c}(t) + \frac{1 - e^{-\Delta\psi(u,t)}}{1 - e^{-\Delta\psi(1,t)}}\, 1_J(t)$$

Proof

Let us first prove that

$$E\left[g(X)\, e^{-u(A_X - A_{X-})}\right]$$

$$= E\int_R g(\tau)\, e^{-u(A_\tau - A_{\tau-})}\, d\left(1 - e^{-A_\tau}\right)$$

$$= \int_R g(\tau)\, e^{\psi(u+1,t-) - \psi(1,t-)}\, d\left(1 - e^{-\psi(u+1,t)}\right)$$

$$- \int_R g(\tau)\, e^{\psi(u,t-) - \psi(1,t-)}\, d\left(1 - e^{-\psi(u,t)}\right)$$

Indeed if $g(\tau) = 1_{(s,t]}(\tau)$ and $\tau_{n,k} = s + \frac{k}{2^n}(t-s)$,

$$E\int_R g(\tau)\, e^{-u(A_\tau - A_{\tau-})}\, d\left(1 - e^{-A_\tau}\right)$$

$$= \lim_{n\to\infty} \sum_{1\leq k\leq 2^n} E\left[e^{-u(A_{\tau_{n,k}} - A_{\tau_{n,k-1}})} \left\{e^{-A_{\tau_{n,k-1}}} - e^{-A_{\tau_{n,k}}}\right\}\right]$$

$$= \lim_{n\to\infty} \sum_{1\leq k\leq 2^n} E\left[e^{-A_{\tau_{n,k-1}}}\right] \left\{E\left[e^{-u(A_{\tau_{n,k}} - A_{\tau_{n,k-1}})}\right]\right.$$

$$\left. - E\left[e^{-(u+1)(A_{\tau_{n,k}} - A_{\tau_{n,k-1}})}\right]\right\}$$

$$= \lim_{n\to\infty} \sum_{1\le k\le 2^n} \frac{E\left[e^{-A_{\tau_{n,k-1}}}\right]}{E\left[e^{-(u+1)A_{\tau_{n,k-1}}}\right]} E\left[e^{-(u+1)A_{\tau_{n,k-1}}} - e^{-(u+1)A_{\tau_{n,k}}}\right]$$

$$- \lim_{n\to\infty} \sum_{1\le k\le 2^n} \frac{E\left[e^{-A_{\tau_{n,k-1}}}\right]}{E\left[e^{-u\,A_{\tau_{n,k-1}}}\right]} E\left[e^{-uA_{\tau_{n,k-1}}} - e^{-uA_{\tau_{n,k}}}\right]$$

$$= \int_R g(\tau)\, e^{\psi(u+1,\tau-) - \psi(1,\tau-)}\, d\left(1 - e^{-\psi(u+1,\tau)}\right)$$

$$- \int_R g(\tau)\, e^{\psi(u,\tau-) - \psi(1,\tau-)}\, d\left(1 - e^{-\psi(u,\tau)}\right)$$

Therefore, it suffices to compute the second member of the right hand side of this last equality. Note first that

$$1_{J^c}(\tau)\, d\left(1 - e^{-\psi(u,\tau)}\right)$$

$$= 1_{J^c}(\tau)\, e^{-\psi(u,\tau)}\, d\psi(u,\tau)$$

$$= 1_{J^c}(\tau)\, e^{-\psi(u,\tau)}\, \varphi(u,\tau)\, d\psi(1,\tau)$$

$$= 1_{J^c}(\tau)\, e^{-\psi(u,\tau) + \psi(1,\tau)}\, \varphi(u,\tau)\, d\left(1-e^{-\psi(1,\tau)}\right)$$

$$= e^{-\psi(u,\tau-) + \psi(1,\tau-)}\, \varphi(u,\tau)\, 1_{J^c}(\tau)\, d\left(1 - e^{-\psi(1,\tau)}\right)$$

On the other hand,

$$1_J(\tau)\, d\left(1 - e^{-\psi(u,\tau)}\right)$$

$$= 1_J(\tau)\, \frac{e^{-\psi(u,\tau-)} - e^{-\psi(u,\tau)}}{e^{-\psi(1,\tau-)} - e^{-\psi(1,\tau)}}\, d\left(1 - e^{-\psi(1,\tau)}\right)$$

$$= e^{-\psi(u,\tau-) + \psi(1,\tau-)}\, \frac{1 - e^{-\Delta\psi(u,\tau)}}{1 - e^{-\Delta\psi(1,\tau)}}\, 1_J(\tau)\, d\left(1 - e^{-\psi(1,\tau)}\right)$$

Hence if we define

$$h(u,\tau) = \varphi(u,\tau)\, 1_{J^c}(\tau) + \frac{1 - e^{-\Delta\psi(u,\tau)}}{1 - e^{-\Delta\psi(1,\tau)}}\, 1_J(\tau)$$

we see that,

$$\int_R g(\tau)\, e^{\psi(u,\tau-) - \psi(1,\tau-)}\, d\left(1 - e^{-\psi(u,\tau)}\right)$$

$$= \int_R g(\tau)\, h(u,\tau)\, d\left(1 - e^{-\psi(1,\tau)}\right)$$

$$= E \int_R g(\tau)\, h(u,\tau)\, d\left(1 - e^{-A_\tau}\right)$$

$$= E[g(X)\, h(u,X)]$$

Therefore

$$E\left[g(X)\,e^{-u(A_X-A_{X-})}\right]$$

$$= E[g(X)\,\{h(u+1,X) - h(u,X)\}]$$

which yields the result.

Acknowledgements

The author is grateful to J.F. Mertens, M. Mouchart, J.P. Raoult and L. Simar for many helpful comments and fructuous conversations on the subject of this paper.

REFERENCES

Blackwell, D. (1973), "Discreteness of Ferguson Selections", *Ann. Stat.* 1, 356-358.

Breiman, L. (1968), *"Probability"*, Addison-Wesley Publishing Company Inc., Reading, Menlo Park and London.

Doksum, K.A. (1974), "Tailfree and Neutral Random Probabilities and their Posterior Distributions", *Ann. Probab.* 2, 183-201.

Fabius, J. (1973), "Two Characterizations of the Dirichlet Distribution", *Ann. Stat.* 1, 583-587.

Ferguson, T.S. (1973), "A Bayesian Analysis of Some Non Parametric Problems", *Ann. Stat.* 1, 209-230.

Ferguson, T.S. (1974), "Prior Distributions on Spaces of Probability Measures", *Ann. Stat.* 2, 615-629.

Ferguson, T.S. and M.J. Klass (1972), "A Representation of Independent Processes without Gaussian Components", *Ann. Math. Stat.* 43, 1634-1643.

Hannum, R.C., Hollander M. and N. Langberg (1981), "Distributional Results for Random Functionals of a Dirichlet Process", *Ann. of Probab.* 9, 665-670.

Itô, K. (1969), *"Stochastic Processes"*, Aarhus, Lecture Note Series n° 16.

James, I.R. and J.E. Mosimann, (1980), "A New Characterization of the Dirichlet Distribution through Neutrality", *Ann. Stat.* 8, 183-189.

Mouchart, M. and L. Simar (1982), "Theory and Applications of Least Squares Approximation in Bayesian Analysis", CORE Discussion Paper n° 8207, University of Louvain-la-Neuve, this volume : chapter 7.

Simar, L. (1982), "A Survey of Bayesian Approaches to Nonparametric Statistics", to appear in *Math. Operationsforsch. Stat., Ser. Stat.*

Wilks, S.S., (1962), *"Mathematical Statistics"*, Wiley, New York.

ROBUST TESTING FOR INDEPENDENT NON IDENTICALLY DISTRIBUTED VARIABLES AND MARKOV CHAINS

by

Lucien Birgé

Université Paris-X Nanterre, U.E.R. de Sciences Economiques
200 Avenue de la République, 92001 Nanterre Cedex (France)

Abstract

In previous papers, it was shown that there exists good tests between two Hellinger balls $\mathcal{P} = B(P_o,r)$ and $\mathcal{Q} = B(Q_o,r)$, given n independent observations. We investigate here the performance of those tests when the true laws of the observations do not belong to those balls; this leads to some robustness result for those tests. The natural extension to non i.i.d. random variables is to define a distance H between product measures by :

$$H^2(P_1 \otimes \dots \otimes P_n, Q_1 \otimes \dots \otimes Q_n) = \sum_{i=1}^{n} h^2(P_i,Q_i)$$

and to test between two balls with this new metric. Reasonable tests are given in this case. We finally present an extension to robust testing for Markov chains and some applications to robust estimation.

Key-words : Neyman-Pearson lemma, Robust tests, Hellinger distance, Non i.i.d. random variables, Markov chains.

A.M.S. 1970 Subject classification : Primary 62G10, Secondary 60J10.

ROBUST TESTING FOR INDEPENDENT NON IDENTICALLY DISTRIBUTED VARIABLES AND MARKOV CHAINS

Lucien BIRGÉ

Université Paris X - NANTERRE

I. INTRODUCTION.

The problem of robust testing has already been considered by many authors from two main points of view. In the first case, one tests two small neighbourhoods of P_n and Q_n, using n i.i.d. variables, and as n increases to infinity, P_n and Q_n get closer to each other while the neighbourhoods schrink to zero. This leads to a purely asymptotic theory and that is the reason why we are not interested in it.

Another way of looking at robust testing is the following. Let us consider two convex neighbourhoods $\mathcal{P}$ and $\mathcal{Q}$ of two probabilities P_0 and Q_0 respectively ; we shall try to find minimax tests of $\mathcal{P}^{\otimes n}$ against $\mathcal{Q}^{\otimes n}$, $\mathcal{P}^{\otimes n}$ being the product of n neighbourhoods, each identical to $\mathcal{P}$. Such a theory has been developped in detail in the well-known papers of Huber and Strassen [9], [10] and [11] for the case where $\mathcal{P}$ and $\mathcal{Q}$ are sets of probabilities under two-alternating capacities and several generalizations were given by Birgé [3], Rieder [19] and Bednarski [1], [2]. Under such hypotheses it is proved that there exists a least favorable pair $(\overline{P_0},\overline{Q_0})$ for testing $\mathcal{P}$ against $\mathcal{Q}$ at any level, such that the likelihood ratio tests of $\overline{P_0}$ against $\overline{Q_0}$ are minimax tests of $\mathcal{P}$ against $\mathcal{Q}$. As a consequence, the likelihood ratio tests of $\bar{P}_0^{\otimes n}$ against $\bar{Q}_0^{\otimes n}$ (with n i.i.d. observations) are also minimax for testing $\mathcal{P}^{\otimes n}$ against $\mathcal{Q}^{\otimes n}$.

Unfortunately, when $\mathcal{P}$ and $\mathcal{Q}$ are not under such capacities, the least favorable pairs depend on the level and it becomes impossible to derive tests of $\mathcal{P}^{\otimes n}$ against $\mathcal{Q}^{\otimes n}$ from such results. In this case there is another solution which has been described in [4]. Instead of looking for minimax tests we content ourselves with finding tests which prove to be asymptotically minimax in the sense of an exponential rate of decrease of errors. This means that both errors of the tests are bounded by exponential functions of n, the coefficients in the exponentials being optimal. This gives us good bounds on the errors for any fixed n. As a consequence we may easily build very simple and explicit tests between two balls for the Hellinger distance h, which is defined by

$$(1.1) \qquad h^2(P,Q) = \frac{1}{2} \int \left(\sqrt{\frac{dP}{d\mu}} - \sqrt{\frac{dQ}{d\mu}}\right)^2 d\mu ,$$

μ being any measure dominating P and Q. Those tests are particularly interesting because they prove very useful in the construction of general estimates, following the methods given by Le Cam [14], [15], [16] or Birgé [5] (see also Dacunha-Castelle [7], Chapter IV).

However, all those properties rely deeply upon the convexity of the neighbourhoods $\mathcal{P}$ and $\mathcal{Q}$. But when we test between such products like $\mathcal{P}^{\otimes n}$ and $\mathcal{Q}^{\otimes n}$ we suppose that all observations $X_1,\dots,X_n$ come from independent trials with respective laws $P_1,\dots,P_n$ all of which belong to the same fixed neighbourhood $\mathcal{P}$ of some given probability P_o. This aspect of robustness does not really take into account the "outliers" or gross errors, but only the small ones. We may also adopt another point of view and suppose that most of the P_i's actually belong to $\mathcal{P}$ but that a small proportion of them are very different from P_o and we do not know to which i's these bad observations correspond (if we knew this, we should just discard them).

Such a conception will lead us to a different way of considering this problem, which has been very seldom used, up to now, probably because it leads to different neighbourhoods which are neither convex sets nor products of convex sets and cannot be treated by the usual tools. Those models were introduced by Le Cam in a somewhat different context (see [14], [15]) but seem to be perfectly suitable for our point of view. Le Cam defines the following distance H between two product measures $P = P_1 \otimes \dots \otimes P_n$ and $Q = Q_1 \otimes \dots \otimes Q_n$:

$$(1.2) \qquad H^2(P,Q) = \sum_{i=1}^{n} h^2(P_i,Q_i).$$

Such a distance defines non-convex neighbourhoods which cannot be expressed as products of Hellinger balls. To see this, let us denote by $B_d(P_o,r)$ the closed ball of center P_o and radius r when we use the distance d (we shall often omit the subscript d when it is clear from the context). It is easily seen that for any $r' \le \frac{r}{\sqrt{n}}$

$$(1.3) \qquad B_H(P_o^{\otimes n}, r) \supset B_h^{\otimes n}(P_o, r')$$

and that $\frac{r}{\sqrt{n}}$ is the largest value for which such an inclusion holds. But obviously $B_H(P_o^{\otimes n}, r)$ is greater than $B_h^{\otimes n}(P_o, \frac{r}{\sqrt{n}})$ and includes sets like $B_h^{\otimes (n-1)}(P_o, \frac{r}{\sqrt{2(n-1)}}) \otimes B_h(P_o, \frac{r}{\sqrt{2}})$. Taking $r = \sqrt{2}$ we see that one of the observations may have an arbitrary law while all other probabilities P_i are within a distance smaller than $\frac{1}{\sqrt{n-1}}$ of P_o. This certainly leads to a non-trivial generalization of the usual notion of robustness.

With this new model, as for the usual ones, we want to find "good" tests between two balls (in H distance). Unfortunately all the usual convexity arguments become useless here. This problem received a partial answer in the papers of Le Cam [14], [15] but only in the case of small balls, of radius smaller than 0.2. According to (1.3) this is equivalent to a generalization of product balls of radius $\frac{1}{5\sqrt{n}}$ for Hellinger distance and proves to be very inadequate for estimation purposes. Actually if we use those tests in the construction of estimators, we get much larger bounds for the risk than we expected, of the type D Log D instead of D , D being usually large (cf. [14], [15]). Our purpose will therefore be to find explicit and reasonnably good tests between the H-balls, independently of their radius. Considerably improved behaviour of the corresponding estimators will follow.

An apparently very different problem is that of tests involving Markov chains. Suppose we are given two chains with respective transition kernels $P(x,.)$ and $Q(x,.)$. How can we define robust tests between those two chains ? Here the problem already lies in the definition of robustness. It will be interesting, as in the case of independent variables, to define robust tests as tests between balls, but this immediately raises a problem because there exists no natural measure of distance between kernels. Starting with some distance d on the space of probabilities, we can easily obtain, for all x, the gauge $d_x(P,Q) = d(P(x,.), Q(x,.))$, but such a gauge depends on x. We could imagine using the supremum or the infinimum with respect to x of such quantities but it is obviously not sensible when $d_x(P,Q)$ is very different from one x to another. A more appealing idea is to use a given probability μ and define $d_\mu(P,Q)$ as $\int d_x(P,Q)\, \mu(dx)$. But how is one to choose μ ?

From another point of view, it would be desirable either for testing or estimation purposes, to use a distance (when it exists) such that, taking n observations of the chain, we could test between P and Q with both errors smaller than

$$\exp[-n K d^2(P,Q)], \quad K > 0, \tag{1.4}$$

K being a fixed constant. Such results are true in the case of n i.i.d. variables if $d = h$. They still hold, as we shall see, for product probabilities with $d = H$ and $n = 1$. To extend to Markov chains some general results found in the case of i.i.d. variables, it would be desirable to find here an analogy of (1.4). We shall see that it is possible if we put a few restrictions on the set of transition kernels that we shall consider. It seems impossible, anyway, to get such results without any hypothesis on the chains.

The purpose of the following paper is to extend to product probabilities (with distance H) and to Markov chains, some results known in the case of i.i.d. variables. Since these new problems have no convexity properties, we shall not try to get minimax tests, but more simply some explicit tests having the property that their errors decrease like exponential functions of the distance between the balls as in (1.4), without looking for optimal values for K. This property of the errors is the essential one for the construction of estimates, whatever the constant K might be. In any case, there is little hope that the methods we are using will lead to very good constants, let alone to the optimal ones.

In the first part of the paper, we shall recall the principal results of [4] concerning tests between Hellinger balls, which will often be used in the sequel. Then we shall prove a technical, but fundamental proposition which will be the key to all subsequent constructions. In Chapter III we shall be able to deduce straightforwardly from this proposition good tests between balls in the distance H. Chapter IV will be devoted to analogous developments in the case of Markov chains under some additional hypotheses; we shall see how these results are connected to the rate of separation of two such chains. As a conclusion, we shall recall how such testing properties could apply to estimation problems and allow the construction of robust estimators with a given rate of convergence in very general cases.

In the sequel, all probability measures are defined on some topological probability space $(E,\mathcal{E})$ which is supposed to be metric and locally compact. This is not a very deep restriction in most applications.

II. PREVIOUS RESULTS and PREREQUISITE.

Let us consider two Hellinger balls $\mathcal{P} = \mathcal{B}_h(P_o,\varepsilon)$ and $\mathcal{Q} = \mathcal{B}_h(Q_o,\eta)$ with $h(P_o,Q_o) \geq \varepsilon + \eta > 0$ and recall that one denotes by $\rho(P_o,Q_o)$ the Hellinger affinity defined by

(2.1) $$\rho(P_o,Q_o) = \int \sqrt{\frac{dP_o}{d\mu}} \sqrt{\frac{dQ_o}{d\mu}}\, d\mu = 1 - h^2(P_o,Q_o),$$

where μ is any positive measure which dominates P_o and Q_o.
We shall normally write $\rho(P_o,Q_o) = \int \sqrt{dP_o\, dQ_o}$ and use such a simplified notation whenever an expression of this type is independent of the dominating measure.

Very simple geometric considerations allow us to prove the existence of a unique pair $(\overline{P_o},\overline{Q_o})$ in $\mathcal{P} \times \mathcal{Q}$ such that

(2.2) $$\rho(\overline{P_o},\overline{Q_o}) = \sup_{P\in\mathcal{P},Q\in\mathcal{Q}} \rho(P,Q) \leq 1 - [h(P_o,Q_o) - \varepsilon - \eta]^2 ,$$

this inequality being strict if $\varepsilon\eta > 0$. $\overline{P_o}$ and $\overline{Q_o}$ may easily be defined by their densities with respect to any measure which dominates P_o+Q_o in the following way :

(2.3) $$\sqrt{d\overline{P_o}} = a \sqrt{dP_o} + a' \sqrt{dQ_o} \;;\; \sqrt{d\overline{Q_o}} = b' \sqrt{dP_o} + b \sqrt{dQ_o} ,$$

a, a', b and b' being the solutions of equations

(2.4) $$\begin{cases} a^2 + a'^2 + 2aa'\,\rho(P_o,Q_o) = 1 \;;\; a + a'\rho(P_o,Q_o) = 1-\varepsilon^2 \\ b^2 + b'^2 + 2bb'\,\rho(P_o,Q_o) = 1 \;;\; b + b'\rho(P_o,Q_o) = 1-\eta^2 . \end{cases}$$

Let us choose a version of the Radon-Nikodym density $\frac{d\overline{Q_o}}{d\overline{P_o}}$ that if $\phi = \sqrt{\frac{d\overline{Q_o}}{d\overline{P_o}}}$ we have

(2.5) $$\underset{\overline{P_o}+\overline{Q_o}}{\text{ess}} \sup \phi = \sup \phi \;;\; \underset{\overline{P_o}+\overline{Q_o}}{\text{ess}} \inf \phi = \inf \phi .$$

Then we can prove the following theorem when E is a locally compact metric space (see [4]) :

<u>THEOREM</u> 1. For any two probability measures P in $\mathcal{P}$ and Q in $\mathcal{Q}$ the following inequalities hold

(2.6) $$\int \phi\, dP \leq \rho(\overline{P_o},\overline{Q_o}) \;;\; \int \phi^{-1}\, dQ \leq \rho(\overline{P_o},\overline{Q_o}).$$

This result has important applications because it shows that ordinary likelihood ratio tests of $\overline{P_0}$ against $\overline{Q_0}$ are also good for testing $P^{\otimes n}$ against $Q^{\otimes n}$. Actually, if we put $P = \bigotimes_{i=1}^{n} P_i$ where all P_i's belong to $\mathcal{P}$, the exponential inequality proves that

$$P\left[\sum_{i=1}^{n} \log \frac{d\overline{Q_0}}{d\overline{P_0}}(x_i) > b\right] \le e^{-\frac{b}{2}} \prod_{i=1}^{n} \left(\int \sqrt{\frac{d\overline{Q_0}}{d\overline{P_0}}}\, dP_i\right)$$

so that using (2.6) we find that the level of the test is smaller than $\exp(-\frac{b}{2})\, \rho^n(\overline{P_0},\overline{Q_0})$ and in the same way that the error of the second kind is bounded by $\exp(\frac{b}{2})\rho^n(\overline{P_0},\overline{Q_0})$. Since we wish to bound both errors at the same time and get symmetrical results we shall fix $b = 0$ and get for both errors the same bound $\exp[-n\, h^2(P,Q)]$ since $h^2(P,Q) = h^2(\overline{P_0},\overline{Q_0}) = 1 - \rho(\overline{P_0},\overline{Q_0})$. Unfortunately this holds only for P in $\mathcal{P}^{\otimes n}$ and Q in $\mathcal{Q}^{\otimes n}$. From the robustness point of view, $P_0^{\otimes n}$ is the ideal model and P the real one, but it is not very easy to fix a precise upper bound to $h(P_0,P_i)$. What will happen if $h(P_0,P_i) > \varepsilon$ for some values of i ? Even if $h(P_i,P_0)$ is only a little larger than ε, Theorem 1 does not give any information though we may easily imagine that some of the observations are a bit worse than we could foresee. So, it would be highly desirable that the tests we use behave well when $h(P_0,P_i)$ is a little larger than ε or even when for a very small number of the i's, the law P_i of X_i is quite different from P_0.

Actually there exists a solution to this problem and we shall give it below in the symmetrical case $(\varepsilon = \eta)$, which is the only one that we shall need in the sequel (but we could generalize this to $\varepsilon \neq \eta$). In this case both equations (2.4) are identical and $a = b$, $a' = b'$. Putting $\rho(P_0,Q_0) = \cos\alpha$ with $0 < \alpha \le \frac{\pi}{2}$, it is easy to see that the solutions will be given by

(2.7) $$a = \frac{\sin(\alpha-\theta)}{\sin\alpha}\ ;\ a' = \frac{\sin\theta}{\sin\alpha}\ ;\ 1-\varepsilon^2 = \rho(P_0,\overline{P_0}) = \rho(Q_0,\overline{Q_0}) = \cos\theta$$

for geometrical reasons. We shall also check that $\rho(\overline{P_0},\overline{Q_0}) = \cos(\alpha - 2\theta)$ so that the balls do not intersect if <u>and</u> only if $0 \le \theta < \frac{\alpha}{2}$. Under those conditions, we shall prove the following proposition which is the key-result for all subsequent proofs. For the sake of simplicity we shall drop the subscripts until the end of this chapter.

PROPOSITION 1. Let P and Q be two probability measures such that $\rho(P,Q) = \cos\alpha$; choose some real number λ with $0 < \lambda < \frac{1}{2}$ and put $r = \sqrt{1-\cos(\lambda\alpha)}$; denote by $(\bar{P},\bar{Q})$ the "least favourable pair" for testing $B(P,r)$ against $B(Q,r)$ which is given by (2.3) and (2.4) and satisfies Theorem 1. Fix $\phi = \sqrt{\frac{d\bar{Q}}{d\bar{P}}}$ so that (2.5) holds. For any probability P' we get the following inequalities

i) if $h(P,P') \le h(P,\bar{P}) = r$ and as a particular case if

(2.8) $$h(P,P') \le \lambda h(P,Q),$$

then

(2.9) $$\int \phi\, dP' \le 1 - h^2(\bar{P},\bar{Q}) \le 1 - (1-2\lambda)^2\, h^2(P,Q).$$

ii) if $h(P,P') > r$ and $A \ge \frac{2}{1-\lambda}$, then

(2.10) $$\int \phi\, dP' \le 1-(1-2\lambda)h^2(P,Q) + \frac{2A(1-2\lambda)}{\lambda(A-2)}h^2(P,P') + \frac{\lambda A(A-2)}{2(1-2\lambda)}\, h^2(\bar{P},\bar{Q}),$$

(2.11) $$\int \phi\, dP' < 1 + \frac{2A(1-2\lambda)}{\lambda(A-2)}\, h^2(P,P')-(1-2\lambda)\ [1-\lambda A(A-2)B(\lambda)]\ h^2(P,Q),$$

with $B(\lambda) = (1-2\lambda)^{-2}\sin^2[\frac{\pi}{4}(1-2\lambda)]$.

Remarks : i) In the course of the proof we shall see that $B(\lambda)$ is an increasing function of λ with range between $\frac{1}{2}$ and $\frac{\pi^2}{16}$.

ii) As a particular case we may choose $A(A-2)B(\lambda) = 2$ (the corresponding values of A belonging to the interval $]3,1+\sqrt{5}]$) and (2.11) becomes

(2.12) $$\int \phi\, dP' \le 1 + \frac{1-2\lambda}{\lambda}\, A^2B(\lambda)h^2(P,P') - (1-2\lambda)^2\, h^2(P,Q).$$

iii) Obviously, from the symmetry of the problem identical results hold if we replace ϕ by ϕ^{-1}, P by Q and $\bar{P}$ by $\bar{Q}$.

Proof of the Proposition : we shall repeatedly use the properties of the function $T(\lambda,\alpha) = \frac{\sin(\lambda\alpha)}{\lambda \sin\alpha}$ for $0 \le \alpha \le \frac{\pi}{2}$, $0 \le \lambda \le 1$ (supposed to be continuous if $\alpha\lambda = 0$). One easily checks that T is decreasing with respect to λ and increasing with respect to α . In particular $T \ge 1$. As a consequence $B(\lambda)$ is an increasing function. From (2.7) we may define $\bar{P}$ and $\bar{Q}$ by

$$\sin\alpha\sqrt{d\bar{P}} = \sin[(1-\lambda)\alpha]\sqrt{dP} + \sin(\lambda\alpha)\sqrt{dQ},$$

$$\sin\alpha\sqrt{d\bar{Q}} = \sin(\lambda\alpha)\sqrt{dP} + \sin[(1-\lambda)\alpha]\sqrt{dQ}$$

and it follows that

$$h^2(P,\bar{P}) = 2\sin^2(\tfrac{\lambda\alpha}{2}) \geq 2\lambda^2\sin^2\tfrac{\alpha}{2} = \lambda^2 h^2(P,Q)$$

so that (2.8) implies $h(P,P') \leq r$. Secondly, from elementary trigonometry we get $h^2(\bar{P},\bar{Q}) = 2\sin^2[\frac{1-2\lambda}{2}\alpha]$ and it entails

$$h^2(\bar{P},\bar{Q}) \leq 4\sin^2[(1-2\lambda)\tfrac{\pi}{4}]\sin^2\tfrac{\alpha}{2} = 2\sin^2[(1-2\lambda)\tfrac{\pi}{4}]\,h^2(P,Q),$$

which proves that (2.10) implies (2.11). Since $h(\bar{P},\bar{Q})$ is also smaller than $(1-2\lambda)\,h(P,Q)$, (2.9) is a simple translation of Theorem 1.

It only remains to prove (2.10) and for this we shall fix $A \geq \frac{2}{1-\lambda}$ and write $P' = f.P + \nu$, ν being orthogonal to P and f choosen in such a way that the set $\{f = +\infty\}$ is exactly the support of ν. Let us write

$E_1 = \{\sqrt{f} > A-1\} \cap \{\phi > 1\}$; $E_2 = \{1 < \sqrt{f} \leq A-1\} \cap \{\phi > 1\}$; $E_3 = \{f < 1\} \cap \{\phi < 1\}$;

$$\Phi = \int \phi\, dP'.$$

Then we have

$$\Phi = \int \phi\, dP + \int (\phi-1)(dP'-dP) \leq \int \phi\, dP + \sum_{i=1}^{3}\int_{E_i}(\phi-1)(dP'-dP).$$

We shall successively bound those four integrals. First

$$\int \phi\, dP = \int \frac{\sin[(1-\lambda)\alpha]\sqrt{dQ}+\sin(\lambda\alpha)\sqrt{dP}}{\sin(\lambda\alpha)\sqrt{dQ}+\sin[(1-\lambda)\alpha]\sqrt{dP}}\, dP = \int \frac{\sin[(1-\lambda)\alpha]\sqrt{\frac{dQ}{dP}}+\sin(\lambda\alpha)}{\sin(\lambda\alpha)\sqrt{\frac{dQ}{dP}}+\sin[(1-\lambda)\alpha]}\, dP$$

with an obvious meaning if $\sqrt{\frac{dQ}{dP}} = +\infty$. The concavity of the function $x \longrightarrow \frac{ax+b}{bx+a}$ for $0 < b \leq a$ and Jensen's inequality imply then

$$\int \phi\, dP \leq \frac{\sin[(1-\lambda)\alpha]\rho(P,Q) + \sin(\lambda\alpha)}{\sin(\lambda\alpha)\rho(P,Q) + \sin[(1-\lambda)\alpha]}\,.$$

Since $\rho(P,Q) = \cos\alpha$ we get

$$\int \phi\, dP \leq \frac{\sin\alpha\cos\alpha\cos(\lambda\alpha)+\sin(\lambda\alpha)[1-\cos^2\alpha]}{\sin\alpha\cos(\lambda\alpha)} = \cos\alpha + \frac{\sin\alpha\sin(\lambda\alpha)}{\cos(\lambda\alpha)}$$

$$\leq 1 - 2\sin^2(\tfrac{\alpha}{2})\left[1 - \frac{tg(\lambda\alpha)}{tg(\frac{\alpha}{2})}\right].$$

But $tg(\lambda\alpha) = tg(2\lambda\frac{\alpha}{2}) \leq 2\lambda\, tg(\frac{\alpha}{2})$ so that finally

(2.13) $$\int \phi \, dP \le 1 - (1-2\lambda)h^2(P,Q).$$

Next we notice that ϕ is bounded from above by $\frac{\sin[(1-\lambda)\alpha]}{\sin(\lambda\alpha)}$. It follows that $\phi - 1 \le \frac{1-2\lambda}{\lambda}$. Since $\frac{f-1}{(\sqrt{f}-1)^2} = \frac{\sqrt{f}+1}{\sqrt{f}-1}$ we see that for $+\infty > \sqrt{f} > A-1$, $(f-1) < \frac{A}{A-2}(\sqrt{f}-1)^2$. Also we clearly have $\int_{\{f=+\infty\}} d\nu = \int_{\{f=+\infty\}} (\sqrt{dP'} - \sqrt{dP})^2$ and finally

(2.14) $$\int_{E_1} (\phi-1)(dP' - dP) \le \frac{1-2\lambda}{\lambda} \frac{A}{A-2} \int_{E_1} (\sqrt{dP'} - \sqrt{dP})^2.$$

If $i = 2$ or 3, the Schwarz inequality entails

$$\left[\int_{E_i} (\phi-1)(dP'-dP)\right]^2 \le \int_{E_i} (\phi-1)^2 dP \int_{E_i} (f-1)^2 dP$$

$$\le \int_{E_i} \left(\sqrt{\frac{d\bar{Q}}{dP}} - \sqrt{\frac{d\bar{P}}{dP}}\right)^2 \frac{dP}{d\bar{P}} dP \int_{E_i} (\sqrt{f}+1)^2(\sqrt{f}-1)^2 dP.$$

On E_2, $\sqrt{f}+1 \le A$ and $\sqrt{\frac{dP}{d\bar{P}}} \le \frac{\sin\alpha}{\sin(\lambda\alpha) + \sin[(1-\lambda)\alpha]}$ since $\frac{dQ}{dP} > 1$; then $\frac{dP}{d\bar{P}} \le 1$ which leads to

(2.15) $$\int_{E_2} (\phi-1)(dP'-dP) \le A\sqrt{\int_{E_2} (\sqrt{d\bar{Q}} - \sqrt{d\bar{P}})^2 \int_{E_2} (\sqrt{dP'} - \sqrt{dP})^2}.$$

In the same way on E_3 we have $\sqrt{f}+1 < 2$ and $\sqrt{\frac{dP}{d\bar{P}}} \le \frac{\sin\alpha}{\sin(1-\lambda)\alpha} \le \frac{1}{1-\lambda}$, and then

(2.16) $$\int_{E_3} (\phi-1)(dP'-dP) \le \frac{2}{1-\lambda}\sqrt{\int_{E_3} (\sqrt{d\bar{Q}} - \sqrt{d\bar{P}})^2 \int_{E_3} (\sqrt{dP'} - \sqrt{dP})^2}.$$

Let us denote

$$I(\Delta) = \int_\Delta (\sqrt{d\bar{Q}} - \sqrt{d\bar{P}})^2, \quad J(\Delta) = \int_\Delta (\sqrt{dP'} - \sqrt{dP})^2;$$

from (2.13), (2.14), (2.15), (2.16) and the inequality $A \ge \frac{2}{1-\lambda}$ we get

(2.17) $$\phi \le 1-(1-2\lambda)h^2(P,Q) + \frac{(1-2\lambda)A}{\lambda(A-2)} J(E_1) + A\left[\sqrt{I(E_2)J(E_2)} + \sqrt{I(E_3)J(E_3)}\right].$$

We also have with $E_4 = E_2 \cup E_3$

(2.18) $$\sqrt{I(E_2)J(E_2)} + \sqrt{I(E_3)J(E_3)} \le \sqrt{I(E_4)J(E_4)}$$

and then noticing that $I(E) = 2h^2(\bar{P},\bar{Q})$ and $J(E) = 2h^2(P,P')$

$$\frac{1-2\lambda}{\lambda(A-2)} J(E_1) + \sqrt{I(E_4)J(E_4)} \le 2\frac{1-2\lambda}{\lambda(A-2)} h^2(P,P') - \frac{1-2\lambda}{\lambda(A-2)} J(E_4) + \sqrt{2h^2(\bar{P},\bar{Q})J(E_4)}.$$

This last expression has a maximal value

$$2\frac{1-2\lambda}{\lambda(A-2)} h^2(P,P') + \frac{\lambda(A-2)}{2(1-2\lambda)} h^2(\bar{P},\bar{Q})$$

for $\sqrt{J(E_4)} = \frac{h(\bar{P},\bar{Q})\lambda(A-2)}{\sqrt{2}\ (1-2\lambda)}$, which we can put into (2.17) using (2.18).
This gives for ϕ the bound (2.10). □

When P, Q, λ and A are fixed the inequalities (2.8) and (2.11) may be written in the form

(2.19) $$\int \phi\, dP' \le 1 - L[h(P,P')]$$

but as L. Le Cam pointed out to me, the function L is not continuous which is neither natural nor aesthetic and leads to poor results when $h(P,P')$ is only a little larger than r. This is why we shall also give an improved version of proposition 1, even if, in all subsequent applications, we shall only use (2.8) and (2.11).

COROLLARY 1 . Let us put $\gamma = 1-2\delta(1-\delta)h^2(P,P')$ and define δ by $\delta = 0$ if $h(P,P') \le r$; $1 - r^2 = \frac{1-(1-\delta)h^2(P,P')}{\sqrt{1-2\delta(1-\delta)h^2(P,P')}}$, $0 < \delta \le 1$ if $h(P,P') > r$. With the hypotheses of Proposition 1 and the conditions $A \ge \frac{2}{1-\lambda}$, $B(\lambda) = (1-2\lambda)^{-2}\sin^2\left[\frac{\pi}{4}(1-2\lambda)\right]$ we get

(2.20) $$\int \phi\, dP' \le \min\{1-h^2(\bar{P},\bar{Q})[\gamma-\delta A\frac{\lambda(A-2)}{2(1-2\lambda)}]+h^2(P,P')\frac{2\delta(1-2\lambda)}{\lambda}[\frac{2(A-1)}{A-2}-\delta];$$
$$1-(1-2\lambda)h^2(P,Q)+\frac{2A(1-2\lambda)}{\lambda(A-2)}h^2(P,P')+\frac{\lambda A(A-2)}{2(1-2\lambda)}h^2(\bar{P},\bar{Q})\}$$

with

(2.21) $$(1-2\lambda)^2h^2(P,Q) \le h^2(\bar{P},\bar{Q}) \le 2B(\lambda)(1-2\lambda)^2h^2(P,Q).$$

Proof : from (2.8) this is obvious if $\delta = 0$. Suppose $h(P,P') > r$, take for μ any positive measure dominating $P+P'$ and put $p = \frac{dP}{d\mu}$, $p' = \frac{dP'}{d\mu}$. δ being as above we define f in $\mathbb{L}^2(\mu)$ by $f = \delta\sqrt{p} + (1-\delta)\sqrt{p'}$ so that

$$\int f\sqrt{p}\, d\mu = \delta + (1-\delta)\rho(P,P') \quad ; \quad \gamma = \int f^2 d\mu = 1-2\delta(1-\delta)h^2(P,P').$$

Then if $P'' = \gamma^{-1}f^2.\mu$, P'' is a probability measure such that

$$\rho(P,P'') = \gamma^{-1/2}[\delta+(1-\delta)\rho(P,P')] = \gamma^{-1/2}[1-(1-\delta)h^2(P,P')] = 1-r^2$$

from the definition of δ. Considering the proof of proposition one we find using (2.8) :

$$\phi = \int \phi\, f^2 d\mu + \int \phi(p'-f^2)d\mu = \gamma\int dP'' + \int(\phi-1)(p'-f^2)d\mu + 1-\gamma$$
$$\le 1-\gamma\, h^2(\bar{P},\bar{Q}) + \int(\phi-1)(p'-f^2)d\mu .$$

But we know that

$$p'-f^2=(\sqrt{p'}-f)(\sqrt{p'}+f) = \delta(\sqrt{p'}-\sqrt{p})[(2-\delta)\sqrt{p'}+\delta\sqrt{p}] = \delta[(1-\delta)(\sqrt{p'}-\sqrt{p})^2+p'-p]$$

from which we find following the proof of Proposition 1 :

$$\begin{aligned}\phi &\le 1-\gamma h^2(\bar{P},\bar{Q})+\delta(1-\delta)\int(\phi-1)(\sqrt{dP'}-\sqrt{dP})^2+\delta\int(\phi-1)(dP'-dP)\\ &\le 1-\gamma h^2(\bar{P},\bar{Q})+\delta(1-\delta)\frac{2(1-2\lambda)}{\lambda}h^2(P,P')+\delta\Big[\frac{2\Lambda(1-2\lambda)}{\lambda(\Lambda-2)}h^2(P,P')+\frac{\lambda\Lambda(\Lambda-2)}{2(1-2\lambda)}h^2(\bar{P},\bar{Q})\Big].\end{aligned}$$

This inequality together with (2.10) prove (2.20). As to (2.21), it was already proved in Proposition 1. □

Remark. If $h(P,P')$ is very close to r, δ is nearly 0 and γ nearly 1 and the first expression in (2.20) is the smallest one. On the contrary, when $h(P,P')$ is large, δ and γ are nearly one and the second expression becomes the best because $(1-2\lambda) > 2\sin^2[\frac{\pi}{4}(1-2\lambda)]$ which entails, $(1-2\lambda)h^2(P;Q) > h^2(\bar{P},\bar{Q})$. This means that (2.20) is a non-trivial improvement of (2.10).

This corollary gives us another inequality of type (2.19) with a continuous and decreasing function L, $L(x)$ being equal to $h^2(\bar{P},\bar{Q})$ when $x \le r$. It is not difficult to see how such results as (2.19) may apply to testing problems. For testing $P_0^{\otimes n}$ against $Q_0^{\otimes n}$ we fix λ and use the likelihood ratio test of $\bar{P}_0^{\otimes n}$ against $\bar{Q}_0^{\otimes n}$ which means that we choose between P_0 and Q_0 according to the sign of $\sum_{i=1}^{n} \log \frac{d\bar{Q}_0}{d\bar{P}_0}(X_i)$. Suppose that the true probability distribution of X_i is P_i, the level of the test will be

$$\alpha = \bigotimes_{i=1}^{n} P_i\Big[\sum_{i=1}^{n} \log \frac{d\bar{Q}_0}{d\bar{P}_0}(x_i) > 0\Big] \le \prod_{i=1}^{n}\Big[\int\sqrt{\frac{d\bar{Q}_0}{d\bar{P}_0}}\, dP_i\Big]$$

from the exponential inequality, which entails

$$(2.22)\qquad \alpha \le \prod_{i=1}^{n}[1-L(h(P_0,P_i))] \le \exp[-\sum_{i=1}^{n} L(h(P_0,P_i))].$$

Whenever all the P_i's are in $\mathcal{P} = B(P_0,r)$, r being related to λ as in Proposition 1, we shall get from (2.8)

$$(2.23)\qquad \alpha \le \exp[-nh^2(\bar{P}_0,\bar{Q}_0)],$$

but using (2.10) or (2.20) we see that α will not change much if a small number of the P_i's are outside $\mathcal{P}$ — or even if a large number of them are at a distance of P_o larger than r, but not much so (which means that δ is near 0 in (2.20)).

Using (2.21) and (2.23) we find for the errors of the tests between $P^{\otimes n}$ and $Q^{\otimes n}$ the bounds

$$\exp[-n(1-2\lambda)^2 \, h^2(P_o,Q_o)]. \tag{2.24}$$

Those results prove very interesting and important in the construction of some special estimates (see [5] and [14]) and the bound (2.24) is an essential hypothesis for such a construction ; that is why we shall now try to generalize it to the cases of non identically distributed variables and Markov chains.

III. ROBUST TESTING FOR INDEPENDENT NON IDENTICALLY DISTRIBUTED VARIABLES.

We shall consider (following here Le Cam in [14]) on the mesurable space $(E^I, \mathcal{E}^I)$, I being a countable set of indexes, the product probabilities $P = \bigotimes_{i\in I} P_i$ and with any two probabilities P and Q of this type we shall associate a semi-distance (because it may be infinite) $H(P,Q)$ given by

$$H^2(P,Q) = \sum_{i\in I} h^2(P_i,Q_i). \tag{3.1}$$

For testing P against Q in some robust way we shall try to find tests between $\mathcal{P} = B_H(P,r)$ and $\mathcal{Q} = B_H(Q,r)$. For this, let us fix some positive $\lambda \le 0.37$; for any i in I let us define the pair $(\overline{P_i},\overline{Q_i})$ as that given by Proposition 1 in which P and Q are replaced by P_i and Q_i and choose the version of the likelihood ratio $\frac{d\bar{P}_i}{d\bar{Q}_i}$ in order to get (2.5). Given the observation $\{x_i\}_{i\in I}$, we shall choose $\mathcal{P}$ or $\mathcal{Q}$ according as $\sum_{i\in I} \log \frac{d\bar{P}_i}{d\bar{Q}_i}(x_i)$ is positive or negative (there is a probability 0 that this sign should depend of the ordering of the series as we shall see).

<u>THEOREM 2</u>. For any P' in $\mathcal{P}$ and Q' in $\mathcal{Q}$, both

$$P'[\sum_{i\in I} \log \frac{d\overline{Q_i}}{d\overline{P_i}}(x_i) \ge 0] \quad \text{and} \quad Q'[\sum_{i\in I} \log \frac{d\overline{P_i}}{d\overline{Q_i}}(x_i) \ge 0]$$

are less than α with

$$(3.2)\qquad \alpha = \exp[-(1-2\lambda)^2 R^2 + \frac{1-2\lambda}{\lambda} B(\lambda)(\bar{A}(\lambda)-2)(3\bar{A}(\lambda)-2)r^2]$$

where $0 < \lambda \leq 0.37$, $0 < R = H(P,Q) \leq +\infty$, $B(\lambda) = (1-2\lambda)^{-2}\sin^2[\frac{\pi}{4}(1-2\lambda)]$, and $\bar{A}(\lambda)$ is the root of $(x-2)^2(x-1)B(\lambda) = 2$.

<u>Proof</u> : from the relation between $\sqrt{\frac{dQ_i}{dP_i}}$ and $\sqrt{\frac{d\bar{Q}_i}{d\bar{P}_i}}$ (see proof of Proposition 1) the series $\Sigma \log \frac{d\bar{Q}_i}{d\bar{P}_i}$ and $\Sigma \log \frac{dQ_i}{dP_i}$ will have the same behaviour. Moreover, since $H(P,P') < +\infty$, P and P' are not orthogonal and by the zero-one law the convergence (or divergence) of $\Sigma \log \frac{dQ_i}{dP_i}$ under P entails the same behaviour under P'. If $H(P,Q) = +\infty$, P and Q are orthogonal and $\Sigma \log \frac{dQ_i}{dP_i} = -\infty$ which P p.s. gives the result. If $H(P,Q)$ is finite, since $\Sigma \log \frac{d\bar{Q}_i}{d\bar{P}_i}$ has a definite sign under $\bar{P} = \otimes_{i\in I} \bar{P}_i$ (as the likelihood ratio of $\bar{P}$ and $\bar{Q}$), this series also has a definite sign under P (since $H(P,\bar{P}) < +\infty$) and also under P'. Then if $P' = \otimes_{i\in I} P'_i$ we may write

$$(3.3)\qquad \gamma = P'[\sum_{i\in I} \log \frac{d\bar{Q}_i}{d\bar{P}_i}(x_i) \geq 0] \leq \prod_{i\in I} \int \sqrt{\frac{d\bar{Q}_i}{d\bar{P}_i}}\, dP'_i \leq \exp[-\sum_{i\in I} L_i(h(P_i,P'_i))]$$

with

$$(3.4)\begin{cases} L_i(x) = (1-2\lambda)^2 h^2(P_i,Q_i) & \text{if } x \leq h(P_i,Q_i) \\ L_i(x) = -\frac{2A(1-2\lambda)}{\lambda(A-2)} x^2 + (1-2\lambda)[1-\lambda A(A-2)B(\lambda)]h^2(P_i,Q_i) & \text{if not,} \end{cases}$$

using Proposition 1. Let J be the subset of I such that $h(P_i,P'_i) > \lambda h(P_i,Q_i)$ if and only if i belongs to J. Since $\sum_J h^2(P_i,P'_i) \leq r^2$, we get $\sum_J h^2(P_i,Q_i) \leq \lambda^{-2} r^2$ and putting $H(P,Q) = R$

$$\text{Log}\,\gamma \leq -(1-2\lambda)^2 \sum_{I-J} h^2(P_i,Q_i) + \frac{2A(1-2\lambda)}{\lambda(A-2)} \sum_J h^2(P_i,P'_i) - \sum_J h^2(P_i,Q_i)(1-2\lambda)[1-A(A-2)B(\lambda)]$$

$$\leq -(1-2\lambda)^2 R^2 + \frac{2A(1-2\lambda)}{\lambda(A-2)} \sum_J h^2(P_i,P'_i) - \sum_J h^2(P_i,Q_i)(1-2\lambda)[2\lambda - A(A-2)B(\lambda)]$$

or more concisely

$$(3.5)\quad \text{Log}\,\gamma \leq -(1-2\lambda)^2 R^2 + \frac{2(1-2\lambda)}{\lambda}\left[\frac{A}{A-2} \sum_J h^2(P_i,P'_i) + \lambda^2\left(\frac{A(A-2)B(\lambda)}{2} - 1\right)\sum_J h^2(P_i,Q_i)\right].$$

Let us denote by Λ the bracketted term in (3.5). We have to find an upper bound of Λ under the restrictions

$$(3.6)\qquad 0 \leq \lambda^2 \sum_J h^2(P_i,Q_i) < \sum_J h^2(P_i,P'_i) \leq r^2 .$$

Let $A_1 = 1 + \sqrt{1 + \frac{2}{B(\lambda)}}$ be the positive root of equation $A(A-2)B(\lambda) = 2$. From (3.5) and (3.6) it is obvious that we have

$$\Lambda \le \frac{A}{A-2}\, r^2 \quad \text{if} \quad A \le A_1 \tag{3.7}$$

$$\Lambda \le r^2\left[\frac{2}{A-2} + \frac{A(A-2)B(\lambda)}{2}\right] \quad \text{if} \quad A \ge A_1 . \tag{3.8}$$

Let us now try to find the best possible value for A. In (3.7) the optimal choice (giving the minimal bound for Λ) for A is A_1. To study (3.8) we shall put $f(x) = \frac{2}{x-2} + \frac{x(x-2)B(\lambda)}{2}$. We find that $f'(x) = -B(\lambda)\left[\frac{2}{B(\lambda)(x-2)^2} - x + 1\right]$ is an increasing function of x when $x > 2$ and that

$$f'(A_1) = -B(\lambda)\left[\frac{A_1}{A_1-2} - A_1 + 1\right] = -\frac{B(\lambda)}{A_1-2}\left[2 - (A_1-2)^2\right]$$

has the same sign as

$$\left(\sqrt{1 + \frac{2}{B(\lambda)}} - 1\right)^2 - 2 = \frac{2}{B(\lambda)}\left(1 - \sqrt{B^2(\lambda) + 2B(\lambda)}\right)$$

which is negative because $B(\lambda) \ge \frac{1}{2}$. As a consequence, $f(x)$ reaches its minimum for a value $\bar{A}(\lambda)$, which is larger than A_1 and given by equation

$$(\bar{A}(\lambda)-2)^2(\bar{A}(\lambda)-1)B(\lambda) = 2. \tag{3.9}$$

This shows that if we take $A = \bar{A}(\lambda) > A_1$ in (3.8), we shall find for Λ an upper bound which is better than that given by (3.7) with $A = A_1$ and equals

$$r^2[\bar{A}(\lambda) - 2]\left[\frac{2}{(\bar{A}(\lambda)-2)^2} + \frac{\bar{A}(\lambda)B(\lambda)}{2}\right].$$

Using (3.9) we may change it into

$$r^2[\bar{A}(\lambda) - 2]\,[3\bar{A}(\lambda) - 2]\frac{B(\lambda)}{2}$$

and find finally :

$$\operatorname{Log} \gamma \le -(1-2\lambda)^2 R^2 + \frac{1-2}{\lambda} B(\lambda)[\bar{A}(\lambda) - 2][3\bar{A}(\lambda) - 2]\, r^2 .$$

The computations are completely analogous for Q'. It is easy to check that if $\lambda \le 0.37$ the condition $A(\lambda) \ge \frac{2}{1-\lambda}$ is fulfilled if we notice that $\bar{A}(\lambda)$ is a decreasing function of λ and $\bar{A}(0.37) > 3.21$, which completes the proof. □

Using (3.2) we see that there exists an optimal value λ_0 of λ which gives the minimum of α, and that λ_0 only depends on the ratio $\frac{R}{r} = C$. The problem is to maximize with respect to λ the expression

$$(3.10) \qquad K(C,\lambda) = (1-2\lambda)^2 - \frac{1-2\lambda}{\lambda}\frac{F(\lambda)}{C} \; ; \; F(\lambda) = B(\lambda)\,[\bar{A}(\lambda) - 2]\,[3\bar{A}(\lambda) - 2].$$

$F(\lambda)$ has been defined as the minimum with respect to x of $\frac{4}{x-2} + x(x-2)B(\lambda)$ and is therefore an increasing function of $B(\lambda)$ and of λ but it does not vary much (since $F(0) = 5.22$ and $F(0.37) = 5.67$) the same being true for $\bar{A}(\lambda)$ (with $\bar{A}(0) < 3.32$ and $\bar{A}(0.37) > 3.21$). $\bar{A}(\lambda)$ is easily computed by Newton's method and we may then maximize $K(C,\lambda)$ using numerical means. First we note that we need $C^2 > 41$ if we want $K(C,\lambda)$ to be positive, and if we take $C = 7$ and $\lambda = \frac{1}{4}$, this will be the case. Therefore we shall restrict our attention to $C \geq 7$ and it only remains to solve the numerical problem.

For different values of C we have computed approximate values $\lambda(C)$ of the optima together with approximate values of the minimum in (3.10). Those results are summarized in the following corollary.

COROLLARY 2. For different values of C the nearly optimal values $\lambda(C)$ and $\bar{K}(C) = K(C,\lambda(C))$ which are given in the table below satisfy the inequality

$$(3.11) \qquad \alpha \leq \exp[-K(C)\, R^2] \; ,$$

where α is given by theorem 2. As particular cases we find $\alpha \leq \exp\left[-\frac{R^2}{10}\right]$ if $C = 8$, $\alpha \leq \exp\left[-\frac{R^2}{3}\right]$ if $C = 12$ and $\alpha \leq \exp\left[-\frac{R^2}{2}\right]$ if $C = 16.52$

C	$\lambda(C)$	$\bar{K}(C)$
7	0.224	0.02545
8	0.181	0.10396
10	0.135	0.23860
12	0.109	0.34226
16.52	0.075	0.50024
20	0.061	0.57912

As an illustration, we may compare these results to those which we get when we deal with balls in the distance h. Suppose that we are given probabilities P_i and Q_i, $i = 1,\dots,n$ with $h(P_i,Q_i) = a_i > 0$ and $\varepsilon_i = \frac{a_i}{C}$, $C \geq 7$. We know that we can find a test between $\bigotimes_{i=1}^n B_h(P_i,\varepsilon_i)$ and $\bigotimes_{i=1}^n B_h(Q_i,\varepsilon_i)$ with errors smaller than $\exp\left[-\sum_{i=1}^n (a_i-2\varepsilon_i)^2\right]$. If we put it in the framework of the space of product probabilities with the distance H we find that $H^2(\bigotimes_{i=1}^n P_i, \bigotimes_{i=1}^n Q_i) = \sum_{i=1}^n a_i^2$ and $B_H(\bigotimes_{i=1}^n P_i, \sqrt{\sum_{i=1}^n \varepsilon_i^2}) \supset \bigotimes_{i=1}^n B_h(P_i,\varepsilon_i)$ with analogous inclusions for Q_i, these inclusions being obviously strict. Using Corollary 2 we see that we can also test the larger balls (using distance H) with errors smaller than $\exp\left[-\sum_{i=1}^n \bar{K} a_i^2\right]$ where $\bar{K}$ depends on the ratio $C = \frac{a_i}{\varepsilon_i}$. Since $\exp\left[-\sum_{i=1}^n (a_i-2\varepsilon_i)^2\right] = \exp\left[-\sum_{i=1}^n (1-2\frac{\varepsilon_i}{a_i})^2 a_i^2\right]$, the only difference is that we have changed $(1-\frac{2}{C})^2$ into $\bar{K}(C)$, which is somewhat smaller as may be seen from Corollary 2, this loss being the price to pay for enlarging the neighbourhoods of $\bigotimes_{i=1}^n P_i$ and $\bigotimes_{i=1}^n Q_i$.

Remark. We supposed that the set of indexes I is countable which is not really important. Either $R < +\infty$ and there exists a countable subset I' of I such that $P_i \neq Q_i$ if i belongs to I' and for i in $I-I'$ we have $\log \frac{dQ_i}{dP_i} = 0$ which does not change anything, or $R = +\infty$ and there exists a countable subset I' for which $\sum_{i\in I'} h^2(P_i,Q_i) = +\infty$ and we may restrict our attention to this subset and get perfect tests (with errors 0) even with this restriction.

Corollary 2 gives us a result for the general case but in the special case when the centers of the balls are laws of i.i.d. variables, we get a stronger result with a very short proof relying on a simple convexity argument. Unfortunately it does not extend to the general case. We may state the following :

THEOREM 2'. For any two probabilities P and Q on E such that $h(P,Q) = \frac{R}{\sqrt{n}}$ we may test between the balls $B_H(P^{\otimes n},r) = \mathcal{P}$ and $B_H(Q^{\otimes n},r) = \mathcal{Q}$ with both errors smaller than α with

(3.12) $\alpha \le \exp\,[-(R-2r)^2]\ , \quad r < \frac{R}{2}\ .$

Proof : put $\rho(P,Q) = 1 - \frac{R^2}{n} = \cos\alpha$, $r' = \frac{r}{\sqrt{n}} = h(P,\bar{P}) = h(Q,\bar{Q})$, $(\bar{P},\bar{Q})$ being the least favorable pair and $\phi = \sqrt{\frac{d\bar{Q}}{d\bar{P}}}$ as in Proposition 1 and suppose that we shall accept P if and only if $\sum_{i=1}^{n} \log\phi(x_i) < 0$. If $P' = \bigotimes_{i=1}^{n} P'_i$ belongs to P we find that

$$\log P'\left[\sum_{i=1}^{n} \log\ \phi(x_i) \ge 0\right] \le \sum_{i=1}^{n} \log \int \phi(x_i)\,dP'_i(x_i) = \delta\ .$$

Using the convexity of the log this gives

$$\delta \le n\ \log \int \phi(x)\ d\tilde{P}(x), \qquad \tilde{P} = \frac{1}{n}\sum_{i=1}^{n} P'_i\ .$$

But obviously by concavity of the square root

$$\rho(P,\tilde{P}) \ge \frac{1}{n}\sum_{i=1}^{n} \rho(P,P'_i) = 1 - \frac{1}{n}\sum_{i=1}^{n} h^2(P,P'_i) \ge 1 - \frac{r}{n} = 1 - r'^2 = \rho(P,\bar{P})$$

which proves that $\tilde{P}$ belongs to the ball $B_h(P,r')$ and that $\int \phi(x)\,d\tilde{P}(x) \le 1 - h^2(\bar{P},\bar{Q})$ and entails that

$$\delta \le n\ \log[1 - h^2(\bar{P},\bar{Q})] \le -n\ h^2(\bar{P},\bar{Q})\,.$$

Since $h(\bar{P},\bar{Q}) \ge \frac{R}{\sqrt{n}} - 2r'$, this completes the proof for the error of first kind. The other case is identical. □

This theorem shows that when the theoretical model deals with i.i.d. variables we may, in our robustness framework enlarge $B_h^{\otimes n}(P,\frac{r}{\sqrt{n}})$ to $B_H(P^{\otimes n},r)$ without loosing anything.

IV. ROBUST TESTING FOR MARKOV CHAINS.

The problem of testing between Markov chains is clearly quite different from the preceding ones. Given two chains with transition kernels $P(x,.)$ and $Q(x,.)$, there exists no "natural" distance between P and Q. In the same way, if $P^n_{x_0}$ and $Q^n_{x_0}$ denote the laws of the vector $(X_1,\ldots,X_n)$ under $P(x,.)$ and $Q(x,.)$ respectively, starting from $X_0 = x_0$, there is no evaluation of the affinity $\rho(P^{\otimes n}_{x_0}, Q^{\otimes n}_{x_0})$ which is an exponential function of n as in the case of i.i.d. variables. We should like, however, to generalize the preceding results to this new case and find a distance d between transitions having the property that we could test between two balls of respective centers $P(x,.)$ and $Q(x,.)$ with errors smaller than $\exp[-n K d^2(P,Q)]$ when the number of the observations is n. As a consequence we would get

$$\rho(P_{x_o}^{\otimes n}, Q_{x_o}^{\otimes n}) \leq 2 \exp [- \frac{n}{2} K d^2(P,Q)]$$

which implies that we can distinguish between the two chains at exponential rate. To ensure such results whatever the starting point x_o may be we shall have to add a few restricting hypotheses.

Beside this, since we are seeking for a general concept of robustness, we shall consider, around the two stationnary chains P and Q, enlarged neighbourhoods including non-stationnary chains. To make this precise we shall first fix some notation.

By "a chain S" we shall mean the set of transition kernels defining the law of the sequence $X_1,\dots,X_n,\dots$. We shall denote by S_x^n the n-dimensional law of the vector $(X_1,\dots,X_n)$ starting from $X_o = x$. $S_x^{i,1}(.)$ will be the transition kernel $S[X_{i+1} \in . | X_i = x]$ and more generally we shall define the iterated kernels by

(4.1) $$S_x^{i,j}(.) = S[X_{i+j} \in . | X_i = x] \quad i \geq 0,\ j \geq 1.$$

To express our hypothesis we suppose we are given two positive integers k and ℓ and two finite positive measures μ and ν $(\neq 0)$ and consider the set $\mathcal{M}$ of chains satisfying the assumption (**H** below) :

Assumption **H** : For all x in E and every integer $i \geq 0$,

(4.2) $$\mu(.) \leq k^{-1} \sum_{j=1}^{k} S_x^{i,j}(.)$$

(4.3) $$S_x^{i,\ell}(.) \leq \nu(.).$$

To understand the meaning of such an assumption we shall suppose that S is stationnary which means that $S_x^{i,1}$ (and obviously $S_x^{i,j}$) is independent of i so that (**H**) says that for any value of X_o and any A in $\mathcal{E}$

(4.4) $$\mu(\Lambda) \leq k^{-1} \sum_{j=1}^{k} S[X_j \in \Lambda], \quad \nu(\Lambda) \geq S[X_\ell \in \Lambda] ,$$

which expresses a kind of uniformity for the chains in $\mathcal{M}$, at least after a certain number of observations. This assumption is clearly a weakened version of the next one, which was used in [5] to prove much weaker results and which also appeared in Donsker and Varadhan [8] :

Assumption $\mathbb{H}'$: There exists a probability measure π and two positive numbers a, b, $0 < a < b$ such that all measures $S_x^{i,1}$ have densities $s_x^{i,1}$ with respect to π satisfying for any x, i and almost any y

(4.5) $$a \leq s_x^{i,1}(y) \leq b.$$

From ($\mathbb{H}'$) we get ($\mathbb{H}$) with $\mu = a.\pi$, $\nu = b.\pi$, $k = \ell = 1$. However ($\mathbb{H}$) is not as restrictive as ($\mathbb{H}'$) : essentially we only need the lower bound for means of the $s_x^{i,j}$ and not separatly for each one. We shall also notice the following useful consequences of ($\mathbb{H}$) with $i \geq 0$ and $m \geq \ell$:

(4.6) $$S_x^{i,m}(.) \leq \nu(.),$$

(4.7) $$\mu(.) \leq k^{-1} \sum_{j=m}^{m+k-1} S_x^{i,j}(.) \leq \nu(.),$$

which can be deduced from the relation

$$S_x^{i,j}(.) = \int S_y^{\ell,j-\ell+i}(.)\, S_x^{i,\ell-i}(dy) \ ; \ i < \ell < j+i.$$

Suppose now that we are given a finite positive measure π , we shall define the "distance" d_π between two transition kernels L and L' by

(4.8) $$d_\pi^2(L,L') = \int h^2[L(x,.),\, L'(x,.)]\, \pi(dx)$$

and the distance d_π between two Markov chains P and Q of length n as the supremum of the distances of the n transition kernels :

(4.9) $$d_\pi(P,Q) = \sup_{i=0,\, n-1} d\, (P_{.}^{i,1}, Q_{.}^{i,1})$$

so that for stationnary chains the two distances are equal (and independent of n). Actually d_π is not a real distance but only a gauge since we may have $d_\pi(L,L') = 0$ and $L \neq L'$ but this is not important for our purpose.

Using, assumption (II) and this definition of distances we shall now give the construction of tests for Markov chains. Let us consider in $\mathcal{M}$ two stationnary chains P and Q with respective kernels $P_x^{0,1} = P_x$ and $Q_x^{0,1} = Q_x$ such that $d_\mu(P,Q) = R$ and two neighbourhoods $\mathcal{P}$ and $\mathcal{Q}$ of P and Q respectively, given by

$$\mathcal{P} = B_{d_\nu}(P,r), \ \mathcal{Q} = B_{d_\nu}(Q,r),$$

μ and ν being as in (II). In order to test $\mathcal{P}$ against $\mathcal{Q}$, using n observations, we shall suppose that $n \geq m = k+\ell$ and put $N = [\![\frac{n}{m}]\!]$ where $[\![\ \ ]\!]$ denotes the integer part ; we shall only use for testing the variables $X_1,\dots,X_p$ with $p = Nm$.

Let us fix $\lambda \leq 0.37$. Then, for all x, by Proposition 1 we may associate with λ, P_x and Q_x a pair $(\overline{P_x},\overline{Q_x})$ which can be used to test P_x against Q_x in a robust way, by considering a good version $\phi(x,t) = \sqrt{\frac{d\bar{Q}_x}{d\bar{P}_x}(t)}$ of their likelihood ratio. Let us define N independent variables $Y_1,\dots,Y_N$ with the same uniform law U on the finite set $\{1;\dots;k\}$, the Y_j being also independent of the X_i, and consider the following statistic ψ

$$(4.10) \qquad \begin{cases} \psi(X_1,\dots,X_p,Y_1,\dots,Y_N) = \sum_{i=1}^{N} \log \phi(X_{J_i-1}, X_{J_i}), \\ J_i = m(i-1) + \ell + Y_i \ . \end{cases}$$

The (randomized) test of $\mathcal{P}$ against $\mathcal{Q}$ that we shall use rejects $\mathcal{P}$ when $\psi > 0$ and accepts it otherwise. To be more specific, we divide the sample into N parts of m observations each, respecting the order; in each part we throw away the ℓ first observations and draw one at random among the $k = m - \ell$ remaining ones, getting a new sample $X_{J_1}, \dots, X_{J_N}$, which we use to compute the likelihood ratio ψ.

<u>THEOREM 3</u>. Suppose we are under (II), $d_\mu(P,Q) = R$, $\mathcal{P} = B_{d_\nu}(P,r)$, $\mathcal{Q} = B_{d_\nu}(Q,r)$; the randomized test which accepts $\mathcal{P}$ if $\psi \leq 0$ has

both errors smaller than

(4.11) $$\alpha = \exp\{ -[\![\tfrac{n}{\ell+k}]\!] [(1-2\lambda)^2 R^2 - \tfrac{1-2\lambda}{\lambda} F(\lambda) r^2]\}$$

with $F(\lambda)$ as in (3.10) and $\lambda \le 0.37$, whatever the law of X_o may be. If $C \ge 7$ and $R \ge Cr$, α is smaller than $\exp[-[\![\tfrac{n}{\ell+k}]\!]\bar{K}(C)R^2]$ with $\bar{K}(C)$ as in Corollary 2.

Proof : we shall restrict ourselves to the error of the first kind, the problem being symmetrical with respect to P and Q. Consider a point S in P ; we want to find an upper bound for

$$\gamma = S_x^n \otimes U^{\otimes N}[\psi(x_1,\dots,x_p,y_1,\dots,y_N) > 0] .$$

Using an exponential inequality and taking the conditionnal expectation with respect to $X_o,\dots,X_{(N-1)m}$ we find

(4.12)
$$\begin{aligned}\gamma &\le \mathbb{E}[\prod_{i=1}^{N} \phi(X_{J_i-1},X_{J_i})]\\ &\le \mathbb{E}[\prod_{i=1}^{N-1} \phi(X_{J_i-1},X_{J_i})\, \mathbb{E}[\phi(X_{J_N-1},X_{J_N}) \mid X_{(N-1)m}]]\end{aligned}$$

Using the independence of X_i's and Y_i's we get

(4.13)
$$\begin{aligned}\mathbb{E}[\phi(X_{J_N-1},X_{J_N})\mid X_{(N-1)m}] &= \sum_{i=1}^{k}\mathbb{E}[\phi(X_{(N-1)m+\ell+i-1},X_{(N-1)m+\ell+i})\mid X_{(N-1)m}]\\ &= k^{-1}\sum_{i=1}^{k}\int \phi(x,y) S_x^{(N-1)m+\ell+i-1,1}(dy)\, S_{X_{(N-1)m}}^{(N-1)m,\ell+i-1}(dx) .\end{aligned}$$

In order to find bounds for (4.13) we shall simplify the notation and put

$$S_x^{(N-1)m+\ell+i-1,1} = S_x'^{i,1} \quad ; \quad S_{X_{(N-1)m}}^{(N-1)m,\ell+i-1} = \bar{S}^i ;$$

$$F_i = \{x \mid h(S_x'^{i,1},P_x) > \lambda h(Q_x,P_x)\}.$$

From Proposition 1 we deduce that whenever $A \ge \frac{2}{1-\lambda}$,

$$\int \phi(x,y) S_x'^{i,1}(dy) \le \begin{cases} 1-(1-2\lambda)^2 h^2(P_x,Q_x) & \text{if } x \notin F_i \\ 1+\frac{2A(1-2\lambda)}{\lambda(A-2)} h^2(P_x,S_x'^{i,1}) - (1-2\lambda)[1-\lambda A(A-2)B(\lambda)]h^2(P_x,Q_x) & \text{if } x \in F_i .\end{cases}$$

As a consequence we find

$$\begin{aligned}\Gamma_i = \iint \phi(x,y) S_x'^{i,1}(dy)\bar{S}^i(dx) &\le 1 - \int_{E-F_i}(1-2\lambda)^2 h^2(P_x,Q_x)\bar{S}^i(dx)\\ &+ \frac{2A(1-2\lambda)}{\lambda(A-2)}\int_{F_i} h^2(P_x,S_x'^{i,1})\bar{S}^i(dx) - (1-2\lambda)[1-\lambda A(A-2)B(\lambda)]\int_{F_i} h^2(P_x,Q_x)\bar{S}^i(dx).\end{aligned}$$

Let us put
$$\int_E h^2(P_x,Q_x)\bar{S}^i(dx) = R_i^2 \ , \ \int_{F_i} h^2(P_x,S'^{i,1}_x)\bar{S}^i(dx) = r_i^2 \ ;$$
the same computation we performed to get (3.5) gives here
$$\Gamma_i \le 1-(1-2\lambda)^2R_i^2 + \frac{2(1-2\lambda)}{\lambda}\left[\frac{\Lambda}{\Lambda-2}\ r_i^2 + \lambda^2\left(\frac{\Lambda(\Lambda-2)B(\lambda)}{2} - 1\right)\int_{F_i} h^2(P_x,Q_x)\bar{S}^i(dx)\right]$$
Taking $\Lambda = \bar{\Lambda}(\lambda)$ as defined by (3.9) we find, following exactly the computations we made in the proof of Theorem 2,
$$\Gamma_i \le 1 - (1-2\lambda)^2\ R_i^2 + \frac{1-2\lambda}{\lambda}F(\lambda)r_i^2 \ ,$$
$F(\lambda)$ being given by (3.10). Then (4.13) becomes
$$\begin{aligned}(4.14)\qquad \mathbb{E}[\phi(X_{J_N-1},X_{J_N}) \mid X_{(N-1)m}] &= k^{-1}\sum_{i=1}^{k}\Gamma_i \\ &\le 1 - \frac{(1-2\lambda)^2}{k}\sum_{i=1}^{k} R_i^2 + \frac{1-2\lambda}{\lambda k}F(\lambda)\sum_{i=1}^{k} r_i^2 \ .\end{aligned}$$
But from (4.6) and (4.7) we know that whatever $X_{(N-1)m}$ may be,
$$k^{-1}\sum_{i=1}^{k} \bar{S}^i(.) \ge \mu(.) \ ; \ \bar{S}^i(.) \le \nu(.),$$
which implies
$$(4.15)\quad k^{-1}\sum_{i=1}^{k} R_i^2 \ge \int_E h^2(P_x,Q_x)\mu(dx) = R^2 \ ; \ r_i^2 \le \int_E h^2(P_x,S'^{i,1}_x)\nu(dx) \le r^2 .$$
From (4.14) and (4.15) we finally find
$$\mathbb{E}[\phi(X_{J_N-1},X_{J_N}) \mid X_{(N-1)m}] \le \exp[-(1-2\lambda)^2R^2 + \frac{1-2\lambda}{\lambda}\ F(\lambda)r^2],$$
and using (4.12)
$$\gamma \le \mathbb{E}[\prod_{i=1}^{N-1} \phi(X_{J_i-1},X_{J_i})]\exp[-(1-2\lambda)^2\ R^2 + \frac{1-2\lambda}{\lambda}\ F(\lambda)r^2] \ .$$
With successive conditionning we get the desired result (4.11) since $N = [\![\frac{n}{\ell+k}]\!]$. The use of $\bar{K}(C)$ is just as in Corollary 2. □

The following Corollary is an immediate consequence of the Theorem ; we do not need the measure ν anymore because we consider the case $r = 0$, so that we only use (4.2). Taking $\lambda = 0$ we get :

<u>COROLLARY 3</u>. Let P and Q be two stationnary Markov chains, k a positive integer, μ a positive measure such that

$$(4.16)\qquad k\mu \le \inf\{\sum_{j=1}^{k} P_x^{o,j}\ ;\ \sum_{j=1}^{k} Q_x^{o,j}\},\ \forall\ x\in E.$$

We may test P against Q using n observations $X_1,\dots,X_n$, whatever be the law of X_o, with errors smaller than

(4.17) $$\exp[-[\![\frac{n}{k+1}]\!] \int h^2(P_x, Q_x)\mu(dx)].$$

This result gives us an exponentional rate of separation for two Markov chains, if we suppose that they are regular enough to satisfy (4.16) with a non-trivial μ. This is a generalization of the classical inequality

$$\rho(P^{\otimes n}, Q^{\otimes n}) = \rho^n(P,Q) \leq \exp[-n\ h^2(P,Q)].$$

V. A FEW POSSIBLE APPLICATIONS TO ESTIMATION THEORY.

We may consider the preceding theorems as robustness results which try to generalize in different ways those of Huber-Strassen [11] and Birgé [4]. Actually, they have been prompted by very different preoccupations related to estimation problems. It was shown by Le Cam [14] and Birgé [5] that using the metric structure of the parameter space Θ, we could give a construction of estimates the rate of convergence of which is closely related with those metric properties, if we have the additional assumption that there exist tests between the balls in Θ with errors satisfying inequalities such as (1.4). Those bounds on the errors of tests were already known in the case of i.i.d. variables (see [13] and [4]) and we had to generalize them. This being done, we shall be able to develop the same metric theory of estimation for the two cases we have studied in the previous sections. In particular, the results of chapter III lead to an immediate improvement of the results of Le Cam [14] and [15] dealing with the rate of estimation for the distance H.

We shall illustrate the possible applications of the results of chapter III and IV by building robust estimates in the case of independent variables and studying the speed of estimation in the case of Markov chains. Before this, we shall make more precise the results that have been found by Le Cam and Birgé on the speeds of estimation, using the notions introduced in [5]. We first recall that in a metric space (Θ,d), a set N is an x-net if any point of Θ is at a distance not larger than x of one point of N. If N is such an x-net, we shall associate with it a constant $\tilde{d}(N,x)$ which is the infimum of all numbers D such that for any integer j, $j \geq 3$, any ball of radius $2^j x$ does not include more than 2^{jD} points of N. It is clear that when Θ is compact, this number will be finite for a good choice of N. We know the following (see [5]).

THEOREM 4. Suppose we are given a set of probability measures $\{P_\theta\}_{\theta\in\Theta}$ indexed by a metric space (Θ,d) and there exist

i) an x-net N of Θ such that $\tilde{d}(N,x) \leq px^2$, $p > 0$;

ii) two positive constants C and K such that for $r \geq x$ and $d(\theta,\theta') \geq Cr$ we may test between $\mathcal{B}(\theta,r)$ and $\mathcal{B}(\theta',r)$ with both errors smaller than $\exp[-Kd^2(\theta,\theta')]$. Then we can find an estimate $\hat{\theta}$ with values in N and such that

$$\sup_{\theta\in\Theta} \mathbb{E}_\theta [d^2(\theta,\hat{\theta})] \leq M(p,K,C)x^2, \tag{5.1}$$

M depending only on p, K, C and not on x or Θ.

Investigating assumption (i) which is of metric type involves approximation theory ; this lies beyond the scope of the present paper and we shall assume that it is fulfilled for some given N and x with $p = 1$, so that in all that follows $\tilde{d}(N,x) \leq x^2$.

Example 1. The typical case where we may apply theorem 4 is that of n i.i.d. variables. The experiment is given by the set $\{P_\theta^{\otimes n}\}_{\theta\in\Theta}$ with the distance

$$d(\theta,\theta') = \sqrt{n}\, h(P_\theta,P_{\theta'}) = H(P_\theta^{\otimes n}, P_{\theta'}^{\otimes n}).$$

The results of [4] give easily condition (ii) for the products of Hellinger balls and in this case (5.1) becomes

$$\sup_{\theta\in\Theta} \mathbb{E}_\theta[nh^2(P_\theta,P_{\hat{\theta}})] \leq M(1,\tfrac{1}{4},4)x^2. \tag{5.2}$$

Bounds on M may be found in [5].

If we wish to perform robust estimation, we may consider, a new parameter space Θ', which will be as follows : take $k \geq 0.16$ and Θ' as the set of product measures $Q = Q_1 \otimes \dots \otimes Q_n$ such that $H(Q, P^{\otimes n}) \leq kx$ for some θ in Θ (depending on Q). We use the distance H on Θ'. By obvious identifications, N becomes a $(k+1)x$-net in Θ' and it is easy to check that the constant $\tilde{H}(N,(k+1)x)$ satisfies

$$\tilde{H}(N,(k+1)x) \leq x^2[1 + \frac{\ell}{3}] , \quad \ell = [\frac{\log \frac{9(k+1)}{8}}{\log 2}] + 1$$

which for $k \geq 0.16$ gives

$$\tilde{H}(N,(k+1)x) \leq (k+1)^2 \ x^2 .$$

Putting $C = 4$, we see that condition (ii) is fulfilled because of Theorem 2'. We may then build a new estimate $\tilde{\theta}$, with values in N and such that

$$\sup_{\theta' \in \Theta'} \mathbb{E}_{\theta'}[H^2(\theta',\theta)] \leq M(1, \tfrac{1}{4}, 4) \ (k+1)^2 x^2 . \tag{5.3}$$

We do not lose much, since if θ' is in Θ , we have $H^2(\theta',\tilde{\theta}) = nh^2(P_{\theta'},P_{\tilde{\theta}})$ so that (5.3) looks very much like (5.2).

Example 2. Let K be a compact in $\mathbb{R}$ (or $\mathbb{R}^p$) and $\{p_\theta(x,y)\}$ a family of positive measurable functions on K^2 (generally rather smooth) and such that for all x $\int_K p_\theta(x,y)dy = 1$.

We define transition kernels $P_{\theta,x}(.)$ by

$$P_{\theta,x}(A) = \int_K 1_A(y)p_\theta(x,y)dy .$$

Now, let us consider the family $\{P_\theta\}_{\theta \in \Theta}$ of Markov chains which are deduced from those kernels and endow Θ with the distance δ defined by

$$\delta^2(\theta,\theta') = \int_K h^2(P_{\theta,x} , P_{\theta',x})dx .$$

With a few suitable smoothness hypotheses on the functions p_θ we shall find constants a and b, $0 < a < b$ such that assumption (II) will be satisfied for the chains in Θ with $\frac{d\mu}{dx} = a$ and $\frac{d\nu}{dx} = b$. Then we may deduce from Theorem 3 that if

$$\delta(\theta,\theta') = R \leq \frac{Cb}{a} r$$

we are able to find tests between the balls $B(\theta,r)$ and $B(\theta',r)$, using the observations $X_1,\ldots,X_n$ and with errors smaller than $\exp[-[\![\frac{n}{\ell+k}]\!]\bar{K}(C)\ a^2R^2]$, $\bar{K}(C)$ being given by Corollary 2. It follows that, taking $d = \sqrt{n}\,\delta$ and N an adequate net in Θ, the assumptions of theorem 4 will be fulfilled and we shall find a bound on the speed of estimation of θ.

As we already mentionned in Chapter IV, to get (**H**), it is enough to check (**H**') , which is much easier and becomes in our case

$$a \le p_\theta(x,y) \le b \qquad \forall\ x, y, \theta.$$

Under such assumptions we proved in [5] an analogous of theorem 3 using uniform distance between p_θ and $p_{\theta'}$. The result is much stronger in this case because the neighbourhoods that we are able to test are larger and at the same time we have weakened the initial assumptions (**H** intead of **H**') . Also, as in example one and in a very similar way, we may find robust estimates by replacing Θ by a neighbourhood Θ' including non-stationnary chains satisfying (**H**). This extension is straightforward and we leave it to the reader.

More generally, to use theorem 4, we only need $\nu \le c\mu$ with some constant c, in (**H**) ; then we choose for d any distance satisfying $d_\mu \le d \le d_\nu$. Actually those conditions are too restrictive and it is possible to weaken them, using two distances simultaneously and working with refined versions of theorem 4, but it becomes much more complicated and we shall not discuss this point. Anyway, it seems very difficult, without any assumption, to define a satisfactory distance between two Markov chains, leading to results of the type of those we found above.

REFERENCES :

[1] BEDNARSKI T. : (1978) Binary experiments, Minimax tests and two-alternating capacities. Institute of Mathematics, Polish Ac. of Sci., Preprint 131.

[2] BEDNARSKI T. : (1981) Solution of minimax test problems for special capacities. Z. f. Wahrscheinlichkeitstheorie u. verw. G. 58 .

[3] BIRGÉ L. : (1977) Tests minimax robustes. Séminaire de statistique d'Orsay 1974-1975, chapitre VII. Astérisque 43-44.

[4] BIRGE L. : (1979) Sur un théorème de minimax et son application aux tests. To appear in Probability and Mathematical Statistics.

[5] BIRGÉ L. : (1981) Approximation dans les espaces métriques et théorie de l'estimation. To be published.

[6] CHERNOFF H. : (1952) A measure of asymptotic efficiency for tests of a hypothesis based on the sum of observations. Ann. Math. Statistics 23, 493-507.

[7] DACUNHA-CASTELLE D. : (1977) Ecole d'Eté de Probabilités de Saint-Flour VII. Lecture notes in Mathematics 678, Springer-Verlag, Berlin.

[8] DONSKER M.D. and VARADHAN S.R.S. : (1975) Asymptotic evaluation of certain Markov process expectations for large time I. Communications on pure and applied mathematics 28, 1-47.

[9] HUBER P.J. : (1965) A robust version of the probability ratio test. Ann. Math Statistics 36, 1753-1758.

[10] HUBER P.J. : (1969) Théorie de l'inférence statistique robuste. Les Presses de l'Université de Montréal.

[11] HUBER P.J. and STRASSEN V. : (1973) Minimax tests and the Neyman-Pearson lemma for capacities. Ann. of Statistics 1, 251-263.

[12] KRAFT C. : (1955) Some conditions for consistency and uniform consistency of statistical procedures. Univ. of California Publ. in Statistics 1 , 125-142.

[13] LE CAM L. : (1973) Convergence of estimates under dimensionality restrictions. Ann. of Statistics 1, 38-53.

[14] LE CAM L. : (1975) On local and global properties in the theory of asymptotic normality of experiments. Stochastic processes and related topics 1, 13-54, Academic Press.

[15] LE CAM L. : (1979) An inequality concerning Bayes estimates. Preprint.

[16] LE CAM L. : (1980) Asymptotic methods in statistical decision theory. Manuscript.

[17] LEHMAN E.L. : (1959) Testing statistical hypotheses. J.Wiley, New-York .

[18] OSTERREICHER F. : (1978) On the construction of least favourable distributions. Z. f. Wahrscheinlichkeitstheorie u. verw. G. 43, 49-55.

[19] RIEDER H. : (1977) Least favourable pairs for special capacities. Ann. of Statistics 5, 909-921.

[20] RIEDER H. : (1978) A robust asymptotic testing model. Ann. of Statistics 6, 1080-1094.

Lucien BIRGÉ (E.R.A. C.N.R.S. 532)
U.E.R. de Sciences Economiques,
UNIVERSITE PARIS X - NANTERRE,
200, Avenue de la République
F-92001 NANTERRE CEDEX (France)

"ON THE USE OF SOME VARIATION DISTANCE INEQUALITIES TO ESTIMATE THE DIFFERENCE BETWEEN SAMPLE AND PERTURBED SAMPLE"

by

A. Hillion

(Université de Bretagne Occidentale*, FRANCE)

Abstract

Given two possible models for the distribution of a sequence of random variables $(X_n)_{n\geq 1}$ (for the first model, the X_n's are i.i.d; for the second model the distribution of each X_n is slightly perturbed), it is suggested to estimate the variation and Hellinger distances between the two possible distributions of $(X_n, X_{n+1},\ldots,X_{n+m})$. The result is derived from some useful inequalities about variation and Hellinger distances on product spaces.

* Département de Mathématiques et Informatique
6, avenue Victor le Gorgeu, 29283 BREST CEDEX.

Key-words : Variation and Hellinger distances on product spaces, perturbed sample, not i.i.d. random variables.

A.M.S. 1970 Subject classification : Primary 62B15, Secondary 62G35.

ON THE USE OF SOME VARIATION DISTANCE INEQUALITIES TO ESTIMATE THE DIFFERENCE BETWEEN SAMPLE AND PERTURBED SAMPLE.

I SUMMARY

Let us consider two possible models for the distribution of a sequence of random variables $(X_n)_{n\geq 1}$, taking their values in some measurable space $(\mathfrak{X}, \mathcal{B})$.

For the first model (<u>sample</u>), the X_n are independent and identically distributed, the law of X_n being P_θ (where $(P_\theta)_{\theta\in\Theta}$ is a family of probability measures defined on $(\mathfrak{X}, \mathcal{B})$ and Θ is a real interval).

We derive the second model (<u>perturbed sample</u>) from the first one by assuming that the distribution of each X_n is lightly perturbed (within the family $(P_\theta)_{\theta\in\Theta}$) : we are given the <u>perturbation process</u> $(Y_n)_{n\geq 1}$ (a sequence of real random variables) and we suppose that the distribution of $X^{(n,n+m)} = (X_n, \ldots, X_{n+m})$ given that $Y^{(n,n+m)} = (y_n, \ldots, y_{n+m})$ is equal to $\bigotimes_{i=n}^{n+m} P_{\theta+y_i}$.

We denote by $P_\theta^{(n,n+m)}$ the distribution of $X^{(n,n+m)}$ for the first model, by $Q^{(n,n+m)}$ the distribution of $X^{(n,n+m)}$ for the second model and by $\Pi_\theta^{(n,n+m)}$ the distribution of $Y^{(n,n+m)}$.

The aim of this paper is to provide tractable conditions on the perturbation process so that, neglecting the first observations, the second model belongs to some neighbourhood (for variation distance d) of the first one.

In order to estimate $\Delta_{n,n+m} = d(P_\theta^{(n,n+m)}, Q_\theta^{(n,n+m)})$, we shall make use of some variation distance inequalities (of which we produce here a simplified demonstration cf. II).

Supposing that the family $(P_\theta)_{\theta \in \Theta}$ satisfies certain regularity conditions, we prove that the two models are close together if the series $\sum_{n=1}^{+\infty} |Y_n|$ converges (cf. III, 2°, theorem 1, the sharpness of which is shown by an example cf III,3°).

Finally, we resume the proceding study by replacing variation distance by Hellinger distance (cf. IV)

II SOME VARIATION DISTANCE INEQUALITIES

------

1) Definitions

Let $(\mathcal{X}, \mathcal{B})$ be a measurable space, P and Q two probability measures on $(\mathcal{X}, \mathcal{B})$ and p and q the densities of P and Q with respect to μ (a positive, σ-finite measure dominating P and Q). We may define,

. the variation distance between P and Q (denoted by $d(P,Q)$)

by (II. 1) $$d(P,Q) = \sup_{B \in \mathcal{B}} |P(B) - Q(B)|$$

. the lower bound of P and Q (denoted by $(P \wedge Q)$) by

(II.2) $$(P\wedge Q)(B) = \inf_{A \in \mathcal{B}} \{P(B \cap A) + Q(B \cap (\mathcal{X} - A))\}$$

or, equivalently, by

(II . 3) $$\frac{d(P\wedge Q)}{d\mu} = \inf(p,q)$$

. $d(P,Q)$ and $(P\wedge Q)$ are then connected by

(II . 4) $d(P,Q) = 1 - \| P\wedge Q \|$ (where $\|P\wedge Q\| = (P\wedge Q)(\mathcal{X})$).

2) Variation distance on product spaces.

If $P^{(2)}$ and $Q^{(2)}$ are the two possible distributions of a couple of random variables (X,Y) (taking their values in $(\mathcal{X}_1, \mathcal{B}_1) \otimes (\mathcal{X}_2, \mathcal{B}_2)$) and defined by the two marginal distributions of X (P_1 and Q_1) and the two conditional distributions of Y given that $\{X = x\}$, it is interesting to connect

$d(P^{(2)}, Q^{(2)})$ with $d(P_1,Q_1)$ and $d(P_x,Q_x)$ (or, more exactly, to connect $(P^{(2)}\wedge Q^{(2)})$ with $(P_1\wedge Q_1)$ and $(P_x\wedge Q_x)$).

Let us suppose (neglecting, for a while, any question of measurability, see. II, 3°) that P_1 and Q_1 have the densities p_1 and q_1 (with respect to some measure μ_1 on $(\mathcal{X}_1, \mathcal{B}_1)$) and that P_x and Q_x have the densities p_x and q_x (with respect to some measure μ_2 or $(\mathcal{X}_2, \mathcal{B}_2)$), the density of $(P^{(2)}\wedge Q^{(2)})$ (with respect to $\mu_1 \otimes \mu_2$) is then equal to $\inf(p_1p_x, q_1q_x)$.

As for every (x,y) belonging to $\mathcal{X}_1 \times \mathcal{X}_2$,
$\inf(p_1(x)p_x(y), q_1(x)q_x(y) \geq \inf(p_1(x),q_1(x)) \inf(p_x(y),q_x(y))$
and as that last quantity is the density (with respect to $\mu_1 \otimes \mu_2$) of the measure $\nu^{(2)}$ defined on $\mathcal{X}_1 \times \mathcal{X}_2$ by

$$\nu^{(2)}(B_1 \times B_2) = \int_{B_1} (P_x\wedge Q_x)(B_2)\,d(P_1\wedge Q_1)(x) \quad B_1\in\mathcal{B}_1, B_2\in\mathcal{B}_2$$

we obtain

(II.5) $(P^{(2)}\wedge Q^{(2)})(B) \geq \nu^{(2)}(B)$ for every B of $\mathcal{B}_1 \otimes \mathcal{B}_2$.

. It is easy to generalize the previous result to the case of an n-tuple of random variables $X^{(n)}=(X_1,\ldots,X_n)$ (taking its values in $(\mathcal{X}^{(n)}, \mathcal{B}^{(n)}) = \bigotimes_{i=1}^{n} (\mathcal{X}_i, \mathcal{B}_i)$). Denoting by $P^{(n)}$ and $Q^{(n)}$ the distributions of $X^{(n)}$, by $P_x(n-1)$ and $Q_x(n-1)$ the conditional distributions of X_n given that $\{X^{(n-1)}=x^{(n-1)}\}$, we may define, by induction, the measure $\nu^{(n)}$ on $(\mathcal{X}^{(n)}, \mathcal{B}^{(n)})$ by

$$\nu^{(n)}(B^{(n-1)} \times B_n) = \int_{B^{(n-1)}} (P_x(n-1) \wedge Q_x(n-1)(B_n))\,d\nu^{(n-1)}(x^{(n-1)})$$

$$B_n \in \mathcal{B}_n,\ B^{(n-1)} \in \mathcal{B}^{(n-1)}$$

We then obtain the following inequalities :

(II.6) $(P^{(n)}\wedge Q^{(n)})(B) \geq \nu^{(n)}(B)$ for every B of $\mathcal{B}^{(n)}$

(II.7) $\|P^{(n)}\ Q^{(n)}\| \geq \|\nu^{(n)}\|$ (by making $B=\mathcal{X}^{(n)}$ in (II.6)) and

(II.8) $\|P^{(n)} \wedge Q^{(n)}\| \geq \prod_{i=1}^{n} \|P_i \wedge Q_i\|$ (which is a particular

case of (II.7) vhen the random variables X_i are independent: $P^{(n)} = \bigotimes_{i=1}^{n} P_i$).

. Let us note finally that is follows from (II.1) or (II.2) that

(II.9) $\|P^{(n-1)} \wedge Q^{(n-1)}\| \geq \|P^{(n)} \wedge Q^{(n)}\|$

3) Remarks

. The previous demonstration of inequality (II.7) is a straightforward generalization of the proof of inequality (II.8) given by W.Hoeffding and J. Wolfowitz ([7]) (see, also, [3] for another demonstration of (II.8)

. In the general case (when the X_n are not necessarily independent), although densities of the type p_x and q_x may not exist or $x \rightsquigarrow \|P_x \wedge Q_x\|$ need not to be measurable, inequality (II.6) remains valid, under the reasonable assumption that the σ-fields $\mathcal{B}_i (2 \leq i \leq n)$ are countably generated (see [5] for the details of the domonstrations based upon (II.2)).

. Morever, inequality (II.7) is sharp (see [1], [9] and, further III,3°, for particular cases when the X_n are independent ; [2] for a concrete case when the X_n are dependent and [5] for a general study).

III PERTURBED SAMPLE

-----

1) Some regularity assumptions

In order to ensure that the perturbations within the family $(P_\theta)_{\theta \in \Theta}$ are rather continuous, we shall assume that the following regularity assumption is satisfied.

(R-A) For every θ of Θ, there exist two (strictly) positive reals $a(\theta)$ and $H(\theta)$ such that

$$d(P_\theta, P_{\theta+t}) \leqslant |t|\, H(\theta) \quad \text{as soon as } |t| \leqslant a(\theta)$$

or, equivalently (by choosing $a(\theta)$ small enough).

(III.1) $\|P_\theta \wedge P_{\theta+t}\| \geqslant 1 - |t|\, H(\theta) \geqslant 0$ as soon as $|t| \leqslant a(\theta)$.

2) Perturbed sample.

(For simplicity, we take $\theta = 0$, writing, for instance, $P^{(n,n+m)}$ instead of $P_\theta^{(n,n+m)}$, H instead of $H(\theta)$...).

Let us consider the two probability measures defined on $(R^m \times \mathfrak{X}^{(m)}, \mathcal{B}(R^m) \otimes \mathcal{B}^{(m)})$ as follows :

$$S = \Pi^{(n,n+m)} \otimes P^{(n,n+m)}$$

T is the distribution of $(Y^{(n,n+m)}, X^{(n,n+m)})$ for the second model ("perturbed sample").

The two marginal distributions on $\mathfrak{X}^{(m)}$ (for S and T) being respectively $P^{(n,n+m)}$ and $Q^{(n,n+m)}$, it follows from (II.9) that

(III.2) $\|P^{(n,n+m)} \wedge Q^{(n,n+m)}\| \geqslant \|S \wedge T\|$

. Inequality (II.7) allows us to write

(III.3) $$\|S \wedge T\| \geqslant \int_{R^m} \Big\| \Big(\bigotimes_{i=n}^{m} P\Big) \wedge \Big(\bigotimes_{i=n}^{n+m} P_{y_i}\Big)\Big\|\, d\Pi^{(n,n+m)}(y_n,..,y_{n+m})$$

and, finally, using (III.2) and (II.8)

(III.4) $$\|P^{(n,n+m)} \wedge Q^{(n,n+m)}\| \geqslant \int_{R^m} \prod_{i=n}^{n+m} \|P \wedge P_{y_i}\|\, d\Pi^{(n,n+m)}(y_n,..,y_{n+m})$$

. Assumption (R-A) implies, for $|y_i|$ small (say $|y_i| < \varepsilon$),

$$\|P \wedge P_{y_i}\| > (1 - |y_i|\, H) \geqslant e^{-2|y_i| H}$$

whence

$$\text{(III.5)}\quad \prod_{i=n}^{n+m} \| P \wedge P_{y_i} \| \geq \exp\{-2H \sum_{i=n}^{n+m} |y_i|\} \quad \text{if } |y_i| < \varepsilon \ (n \leq i \leq n+m)$$

. As, for every $y^{(n,n+m)}$ belonging to $A(\varepsilon) = \{(y_n,\ldots,y_{n+m}) : \sum_{i=n}^{n+m} |y_i| < \varepsilon\}$, we have

$\exp\{-2H \sum_{i=n}^{n+m} |y_i|\} \geq e^{-2\varepsilon H}$, we get, from (III.4) and (III.5),

$$\text{(III.6)}\quad \| P^{(n,n+m)} \wedge Q^{(n,n+m)} \| \geq e^{-2\varepsilon H}\, \Pi\{\sum_{i=n}^{n+m} |Y_i| < \varepsilon\}$$

, which may be written

$$d(P^{(n,n+m)}, Q^{(n,n+m)}) \leq 1 - e^{-2\varepsilon H}\, \Pi\{\sum_{i=n}^{n+m} |Y_i| < \varepsilon\}$$

or, by letting m increase indefinitely,

$$\text{(III.7)}\quad d(P^{(n,+\infty)}, Q^{(n,+\infty)}) \leq 1 - e^{-2\varepsilon H}\, \Pi\{\sum_{i=n}^{+\infty} |Y_i| < \varepsilon\}$$

(where $d(P^{(n,+\infty)}, Q^{(n,+\infty)})$ is the variation distance between the two possible distributions of the process $(X_i)_{i \geq n}$).

. Inequality (III.7) etablishes the following theorem

Theorem 1

If $\sum_{n=1}^{+\infty} Y_n$ converges in probability,

$$\lim_{n \to +\infty} d(P^{(n,+\infty)}, Q^{(n,+\infty)}) = 0$$

3) **Example**

The following example shows that the convergence of $\sum_{n=1}^{+\infty} |Y_n|$ may be a necessary and sufficient condition so that $\lim_{n \to +\infty} d(P^{(n,+\infty)}, Q^{(n,+\infty)}) = 0$.

. Let us suppose that

1) $\Theta = [0,1]$, $P_\theta = \mathcal{U}([\theta\ ,\ 1+\theta])$ (uniform distribution on $[\theta, 1+\theta]$). As $\|P_o \wedge P_t\| = 1 - t$, assumption R-A is satisfied.

2) The perturbation random variables $(Y_n)_{n \geq 1}$ are independent, the distribution of Y_i being $\mathcal{U}([0,\varepsilon_i])$ $(0<\varepsilon_i<1)$.

Thus, for the two possible models, the X_i are independent : for the first model, the distribution of each X_i is P_o ; for the second model, the density of Q_i, the distribution of X_i, is found to be

$$q_i = \begin{cases} \dfrac{x}{\varepsilon_i} & \text{if } 0 \leq x \leq \varepsilon_i \\ 1 & \text{if } \varepsilon_i \leq x \leq 1 \\ \dfrac{1+\varepsilon_i - x}{\varepsilon_i} & \text{if } 1 \leq x \leq 1+\varepsilon_i \end{cases}$$

. To compute $d(P_o, Q_i)$, we make use of the classical formula $d(P_o, Q_i) = P_o(H_i) - Q_i(H_i)$ (where $H_i = \{p_o \geq q_i\}$. Here $H_i = [0,1]$ and $P_o\ \{p_o \geq q_i\} = 1$, whence

$$\|P^{(n,n+m)} \wedge Q^{(n,n+m)}\| = \prod_{i=n}^{n+m} \|P_o \wedge Q_i\| = \prod_{i=n}^{n+m} \left(1 - \frac{\varepsilon_i}{2}\right)$$

. Thus $\lim_{n\to+\infty} d(P^{(n,+\infty)}, Q^{(n,+\infty)}) = 0$ if and only if $\sum_{n=1}^{+\infty} \varepsilon_n < +\infty$ (which is a necessary and sufficient condition of stochastic convergence of $\sum_{n=1}^{+\infty} |Y_n|$, since the Y_n are independent).

4) <u>Remark</u>

It is easy to get uniform results in theorem 1 (i-e $\lim_{n\to+\infty} \left[\sup_\theta d(P_\theta^{(n,+\infty)}, Q_\theta^{(n,+\infty)})\right] = 0$) by assuming that the following uniform conditions are satisfied :

. The series $\sum_{n=1}^{+\infty} |Y_n|^r$ converges uniformly (in probability) over Θ

. There exist three (strictly) positive reals, a, r and H such that $\|P_\theta \wedge P_{\theta+t}\| \geqslant 1-|t|^r H$ for every $|t| < a$ and every θ of Θ.

IV HELLINGER DISTANCE

Although the previous example (III.3) shows that theorem 1 is sharp under assumption (R-A), it is possible to prove that, under stronger regularity assumptions (see IV.2), $\lim_{n\to+\infty} d(P^{(n,+\infty)}, Q^{(n,+\infty)}) = 0$ as soon as $\sum_{n=1}^{+\infty} Y_n^2 < +\infty$ in probability (cf. IV,3). Let us recall a few properties of Hellinger distance which will be useful for deriving theorem 2 (IV,3°).

1) Hellinger distance on product spaces

. We define

. measure $(PQ)^{\frac{1}{2}}$ by its density (with respect to μ) $p^{\frac{1}{2}} q^{\frac{1}{2}}$.

. the affinity between P and Q (denoted by $\rho(P,Q)$) by

$$\text{(IV.1)} \quad \rho(P,Q) = \|(PQ)^{\frac{1}{2}}\| = \int_{\mathcal{X}} p^{\frac{1}{2}} q^{\frac{1}{2}} \, d\mu$$

. the Hellinger distance between P and Q (denoted by $D(P,Q)$) by $D^2(P,Q) = 1 - \rho(P,Q)$.

. variation distance and Hellinger distance are connected as follows

$$\text{(IV.2)} \quad \sqrt{2}\, D \geqslant d \geqslant D^2$$

. In order to connect $D(P^{(n)},Q^{(n)})$ with expressions $D(P_x(n-1), Q_x(n-1))$ ($P^{(n)}$ and $Q^{(n)}$ being defined as in II.2), it is useful to define, by induction, the measure $\lambda^{(n)}$ on

$(\mathfrak{X}^{(n)}, \mathcal{B}^{(n)})$ by

$$\lambda^{(n)}(B^{(n-1)} \times B_n) = \int_{B^{(n-1)}} (P_{x(n-1)}\, Q_{x(n-1)})^{\frac{1}{2}}(B_n)\, d\lambda^{(n-1)}(x^{(n-1)})$$

$B_n \in \mathcal{B}_n$ and $B^{(n-1)} \in \mathcal{B}^{(n-1)}$

and $\lambda^{(1)} = (P_1 Q_1)^{\frac{1}{2}}$.

. Thus we get (see [5] , for some study in the general case).

. $(P^{(n)}\, Q^{(n)})^{\frac{1}{2}} = \lambda^{(n)}$

(IV.4) . $\rho(P^{(n)}, Q^{(n)}) = \|\lambda^{(n)}\|$ which may be written

(IV.5) . $\rho(P^{(n)}, Q^{(n)}) = \prod_{i=1}^{n} \rho(P_i, Q_i)$ when the random variables are independent.

2) Regularity assumption

It may be shown (cf. [6]), that, if the family $(P_\theta)_{\theta \in \Theta}$ is sufficiently regular (if, for instance, P_θ being absolutely continuous with respect to P_o, application $\theta \mapsto \left(\frac{dP_\theta}{dP_o}\right)^{1/2}$ has an asymptotic expansion in $L^2(\mathfrak{X}, \mathcal{B}, P_o)$ about $\theta = 0$), there exist two (strictly) positive reals such that

(IV.6) $\lim_{t \to 0} \frac{d(P_o, P_t)}{|t|} = a$ and $\lim_{t \to 0} \frac{D(P_o, P_t)}{|t|} = b.$

For such a family, we have

$\|P_o \wedge P_t\| > 1 - |t|\, H$ (i-e assumption (R-A))

but too

$\rho(P_o, P_t) \geqslant 1 - t^2 K.$

3) Theorem

Let us suppose that the following regularity assumption is satisfied (R-A') . For every θ of Θ, there exist two (strictly) positive reals $a(\theta)$ and $H(\theta)$ such that

$\rho(P_\theta, P_{\theta+t}) \geq 1 - t^2 K(\theta)$ as soon as $|t| \leq a(\theta)$.

Thus we get

Theorem 2

If $\sum_{n=1}^{+\infty} |Y_n|^2$ converges in probability,

$$\lim_{n \to +\infty} d(P^{(n,+\infty)}, Q^{(n,+\infty)}) = 0.$$

Proof

The demonstration is parallel to that developped in III.2, putting $(PQ)^{\frac{1}{2}}$ in place of $(P \wedge Q)$.

Assumption (R-A') will lead to

$$\rho(P^{(n,n+m)}, Q^{(n,n+m)}) \geq e^{-2\varepsilon K} \Pi \{ \sum_{i=n}^{n+m} Y_i^2 < \varepsilon \}$$

4) **Example**

The following example shows that the convergence of $\sum_{n=1}^{+\infty} |Y_n|^2$ may be a necessary and sufficient condition so that $\lim_{n \to +\infty} d(P^{(n,n+\infty)}, Q^{(n,n+\infty)}) = 0$.

. Let us suppose that

1) $\Theta = \mathbb{R}$, $P_\theta = \mathcal{N}(\theta, 1)$ (gaussian distribution with mean θ and variance 1) As $\rho(P_0, P_\theta) = e^{-\frac{\theta^2}{8}}$, assumption R-A' is satisfied.

2) The perturbation random variables $(Y_n)_{n \geq 1}$ are independent, the distribution of Y_i being $\mathcal{N}(\varepsilon_i, \varepsilon_i^2)$ (with $\lim_{i \to +\infty} \varepsilon_i = 0$).

. Thus, for the second model, the distribution of X_i, Q_i is found to be equal to $\mathcal{N}(\varepsilon_i, 1 + \varepsilon_i^2)$

. A routine computation of affinity (cf, for instance, [8]) gives

$$\rho(P_o,Q_i) = \frac{2^{\frac{1}{2}}(1+\varepsilon_i^2)^{\frac{1}{4}}}{(2+\varepsilon_i^2)^{\frac{1}{2}}} \quad \exp\left(\frac{-\ \varepsilon_i^2}{4(2+\varepsilon_i^2)}\right)$$

As $\rho(P_o,Q_i) = 1 - \frac{\varepsilon_i^2}{8} + o(\varepsilon_i^2)$ and

$$\rho(P^{(n,n+m)},\ Q^{(n,n+m)}) = \prod_{i=n}^{n+m} \rho(P_o,Q_i)$$

$\lim_{n\to+\infty} D(P^{(n,+\infty)},Q^{(n,+\infty)}) = 0$ if and only if $\sum_{n=1}^{+\infty} \varepsilon_n^2<+\infty$ (which is a necessary and sufficient condition of stochastic convergence of $\sum_{n=1}^{+\infty} |Y_n|^2$ since the Y_n are independent).

References

[1] S. ABOU-JAOUDE "*Sur un théorème de non-existence d'estimateurs convergeant en probabilité*" Comptes-rendus Acad. Sc. (France), série A, t. 278, p. 1445-1448 (1974).

[2] D.A. FREEDMAN "*A remark on the difference between sampling with and without replacement*" J. Amer. Statist. Ass., Vol 72, p. 681, (1977).

[3] J. GEFFROY "*Distance en variation des lois-produits*" Séminaire de Statistique mathématique, Université Pierre et Marie Curie, Paris (1974-1975), unpublished.

[4] J. GEFFROY "*Asymptotic separation of distributions and convergence properties of tests and estimators*" in "*Asymptotic theory of statistical tests and estimation*" Chakravarti édit., Academic Press 1980, p. 159-178 (1979).

[5] A. HILLION "*Quelques questions de distances de lois de probabilité*" Pub. Inst. Stat. Univ. de Paris, vol. XXIII, fasc 3-4, p. 37-88, (1978).

[6] A. HILLION "*Etude, du point de vue de la séparation asymptotique, de certaines questions de distance de lois de probabilité sur des espaces-produits, d'estimation ponctuelle et par régions de confiance*" Thèse Université Pierre et Marie Curie, Paris (1980).

[7] W. HOEFFDING, J. WOLFOWITZ *"Distinguishability of sets of distributions"* Ann. Math. Statist., vol. 29, p. 700-718 (1958).

[8] K. MATUSITA *"A distance and related statistics in multivariate analysis"* in *"Multivariate analysis"* P.R. Krishnaiah edit., Academic Press p. 187-200 (1966).

[9] R. MOCHE *"Décantation et séparation asymptotique uniformes; tests et estimateurs convergents, dans le cas d'observations indépendantes, équidistribuées ou non"* Thèse (première partie) Université des Sciences et Techniques de Lille (1977).

A CONTRIBUTION TO ROBUST PRINCIPAL COMPONENT ANALYSIS

Jacques Benasseni
CRIG - USTL
Montpellier - FRANCE

ABSTRACT

In the situation where a $p \times n$ data matrix is formed by a sample $\{X^1, \ldots, X^n\}$ from a p variate density of the form $f_{t,V}(x) = (\det V)^{-\frac{1}{2}} h\{(x-t)' V^{-1}(x-t)\}$, Maronna (2) proposed a robust estimation of the location vector t and scatter matrix V by means of "M-estimators" defined as solutions of the system :

$$\sum_{i=1}^{n} u_1(d_i)(X^i-t) = 0 \quad \text{and} \quad \sum_{i=1}^{n} u_2(d_i^2)(X^i-t)(X^i-t)' = V$$

where $d_i^2 = (X^i-t)' V^{-1}(X^i-t)$ and u_1, u_2 are non-negative, non-increasing, continuous functions. Then a natural way to get a robust principal component analysis can be to use t and V. Unfortunately V is not the usual scatter matrix since t is computed with weights $p_{t,i} = u_1(d_i) / \sum_{i=1}^{n} u_1(d_i)$ and V with $p_{V,i} = u_2(d_i^2)$.

In this paper we try to explain the meaning of an analysis based on such t and V using Pages and Caillez's duality diagram commonly used in data analysis. For further information one can refer for example to Y. Escoufier and P. Robert (1979).

Key-words : Principal component analysis, M-estimators, Robustness analysis, Data analysis

AMS/MOS : Primary 62H25, Secondary 62G35

1. SOME NOTATION AND RESULTS IN THE CLASSICAL SITUATION WITH ONLY ONE FORM OF WEIGHTS

X being the data matrix giving the values of the p variables for the n observations, let Q be the pXp positive definite (or semi-definite) matrix used to compute distances between the observations in $E = R^p$ and let $D = \text{diag}\,(p_{V,i})$ be the diagonal weight matrix of the observations that plays the role of a metric between variables in $F = R^n$.

The situation is thus completely defined by the triplet (X,Q,D) also called "study (X,Q,D)".

The origin taken in E is $g_D = \sum_{i=1}^{n} p_{V,i}\, X^i$ (since $I_a = \sum_{i=1}^{n} p_{V,i}\, \|X^i - a\|_Q$ is minimized by $a = g_D$) so that in fact we use $X_1 = X(I - D\,e\,e')$ instead of X (with $e = (1, \ldots, 1)'$ in $\mathbb{R}^n$ and I the identity matrix in $\mathbb{R}^n$).

Observations can be represented by the $(X_1)^i$ in E with the metric Q or in F^* (dual space of F) by the f_i^* with metric $W_1 = X_1' Q X_1$ so that distances computed in E and F^* are the same ($(f_i^*)_{i=1,\ldots,n}$ being the canonical basis of F^*). Similarly variables can be represented in F with metric D or in E^* with metric $V_1 = X_1 D X_1'$.

The matrix associated with the corresponding mapping from F^* onto E (resp. from E^* onto F) is obviously X (resp. X').

The situation is then summarized in the following "duality diagram" :

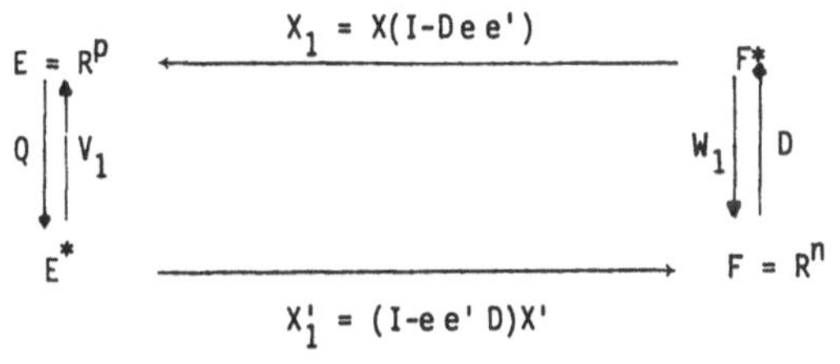

The study (X_1,Q,D) has the following well known properties :

- V_1 is the usual scatter matrix;
- $I_{g_D} = \text{Tr}(X_1' Q X_1 D) = \text{Tr}(X_1 D X_1' Q) = \text{Tr}(W_1 D) = \text{Tr}(V_1 Q)$;
- $W_1 De = 0$ (since $X_1 De = 0$);
- If $W_1 D U_i = \lambda_i U_i$ and $W_1 D U_j = \lambda_j U_j$ with $\lambda_i \neq \lambda_j$ ($U_i, U_j \in \mathbb{R}^n$, $\lambda_i, \lambda_j \in \mathbb{R}$), then $U_i' D U_j = 0$;
- $V_1 Q$ and $W_1 D$ have the same eigenvalues λ_i. Let Γ be the diagonal eigenvalue matrix.

The principal component analysis has the following properties :

- The principal axes in E, columns of the pXq matrix Z $(q < p)$ are the eigenvectors of V_1Q with $Z'QZ = \Gamma$.
- The principal factors in E^*, columns of $L = QZ\Gamma^{-\frac{1}{2}}$ are the eigenvectors of QV_1 with $L'Q^{-1}L = I$.
- Lines of $Y = L'X$ (qXn matrix), elements of $\mathbb{R}^n$, with $YDY' = \Gamma$ are the principal components.

2. THE ROBUST PRINCIPAL COMPONENT ANALYSIS

Let Δ be the diagonal $p_{t,i}$ matrix. The analysis based on V and t as defined by Maronna (1976) is summarized in the "new duality diagram" :

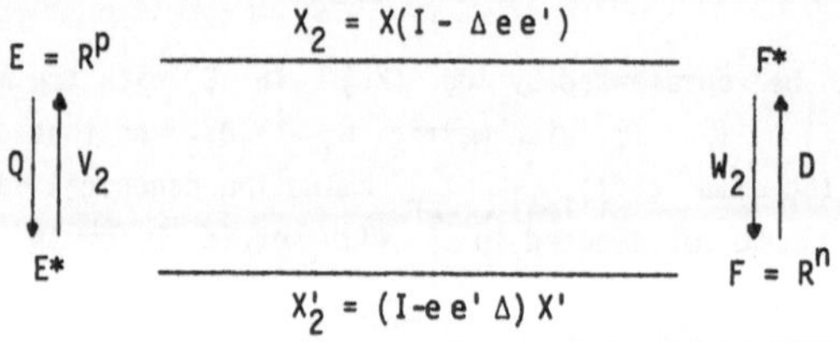

with $V_2 = X_2 DX_2'$ and $W_2 = X_2' QX_2$.

To study properties of this new diagram we shall use the two following simple results :

Lemma 1 : If A and B are two diagonal weight matrices then :

$$(I-Aee')(I-Bee') = (I-Bee').$$

Lemma 2 : If A is a diagonal weight matrix and S a positive semi-definite matrix, then

$$\hat{S} = (I-ee'A)\,S\,(I-Aee')$$

is positive semi-definite and $\hat{S}Ae = 0$.

Therefore :

- $W_2 = (I-ee'\Delta)W_1(I-\Delta ee')$ and W_2 is positive semi-definite.
- If $g_\Delta = \sum_{i=1}^{n} p_{t,i}\, X^i$: $I_{g_\Delta} = Tr(W_2D) = Tr(V_2Q)$.
- $W_2\Delta e = 0$ (because $X_2\Delta e = 0$).

$\cdot\ I_{g_\Delta} = Tr(V_2Q) = I_{g_D} + \|g_\Delta - g_D\|^2_Q = Tr(V_1Q) + \|g_\Delta - g_D\|^2_Q$

$\quad = Tr(W_2D) = Tr(W_1D) + \|g_\Delta - g_D\|^2_Q$

$\cdot\ V_2^{kl} = \sum_{i=1}^{n} pv,i\ (X_k^i - (g_D)_k)\ (X_l^i - (g_D)_l).$

To study the principal component analysis of (X_2,Q,D) we have found a metric M in F such that results of the analysis of (X_2,Q,D) are "similar" to those of (X_1,Q,M).

Let us consider $M = (I-\Delta e e')D\ (I-e e'\Delta)$.

Theorem 1 : Studies (X_2,Q,D) and (X_1,Q,M) have the same principal axes and factors. The result is quite obvious since : $V_2 = X_2DX_2' = X_1MX_1' = V_3$.

Theorem 2 : If Y_2 and Y_3 are matrices whose lines are principal components of (X_2,Q,D) and (X_1,Q,M) respectively, we have :

$$Y_2 = Y_3(I - \Delta e e').$$

Proof : If Λ and Λ^* are diagonal eigenvalues matrices of $X_2'QX_2D$ and $X_1'QX_1M$ respectively, we have by definition of Y_2 and Y_3

(1) $(I-e e'\Delta)\, X'\, QX(I-\Delta e e')\, DY_2' = Y_2'\Lambda$ with $Y_2DY_2' = \Lambda$

(2) $(I-e e' D)\, X'\, QX(I-De e')\, (I-\Delta e e')\, D(I-e e'\Delta)Y_3' = Y_3'\Lambda$ with $Y_3MY_3' = \Lambda^*$.

Multiplying (2) on the left by $(I-e e'\Delta)$ and applying lemma 1 we then have :

(3) $(I-e e'\Delta)\, X'\, QX(I-\Delta e e')D\, (I-e e'\Delta)\, Y_3' = (I-e e'\Delta)Y_3'\Lambda^*$.

Comparison of (1) and (3) shows that $\Lambda = \Lambda^*$ and $Y_2 = Y_3(I-\Delta e e')$

So we see that studies (X_1,Q,M) and (X_2,Q,D) give the same representation of variables and a translated representation of observations. Nevertheless, one could notice that studies are different from the point of view of the cross-product between the characteristic operators W_2D and W_1M since we have

$$Rv\ (W_2D\, ,W_1M) = \frac{Tr(X_2'\,QX_2\,DX_1'\,QX_1\,M)}{\sqrt{Tr(X_2'\,QX_2D)^2\,Tr(X_1'\,QX_1\,M)^2}} = \frac{Tr\{(I-e e'\,D)\,(W_2D)^2\}}{Tr\,(W_2D)^2} \neq 1.$$

Theorem 3 : Principal components of (X_2,Q,D) are such that

$$Y_2 = Y_2(I-\Delta e e') \quad \text{and} \quad Y_2 D Y_2' = \Lambda.$$

Remark : It does not seem possible to find a metric L in E such that results of the principal component analysis of (X_1,L,D) are similar to those of (X_2,Q,D).

3. CONCLUSIONS

The robust principal component analysis presented gives the same results as the usual analysis with a non-diagonal metric on the space of variables F.

In a more general context than robustness, it is possible by introducing a second weight matrix Δ, to obtain principal components with a particular origin which is not the usual g_D. One can even take a particular observation i_o as origin by putting $p_{t,i_o} = 1$ and $p_{t,i} = 0$ for i different from i_o (with eventually $p_{V,i} = 0$ if i_o is an extra observation). Conclusions drawn from factorial plans are then entirely linked to the value of $\|g_D-g_\Delta\|_Q$ in each of them.

4. REFERENCES

Escoufier, Y. and P. Robert (1979), "Optimizing Rv Coefficient", *Optimizing Methods in Statistics,* edited by Jagdish S. Rustagi, Academic Press.

Maronna, R.A. (1976), "Robust M-Estimates of Multivariate Location and Scatter, *Ann. of Statist.*, 4,51-67.

FROM NON PARAMETRIC REGRESSION TO NON PARAMETRIC PREDICTION : SURVEY OF THE MEAN SQUARE ERROR AND ORIGINAL RESULTS ON THE PREDICTOGRAM

Gérard Collomb

Université Paul Sabatier

Laboratoire de statistique et probabilités

Toulouse

ABSTRACT

This paper is made up of two parts which both deal with the mean square error (m.s.e.) of non parametric estimators (n.p.e.). First, we review a number of results on the m.s.e. of various n.p.e. of regression (paragraph 2) and a few results on non parametric prediction (paragraph 3). Secondly (paragraph 4) we present our own results on the predictogram, which is considered as a n.p.e. of the prediction function. A necessary and sufficient condition for L_2 consistency is obtained and the rate of convergence of the predictogram is investigated.

These different non parametric methods are studied from an asymptotic statistics point of view and in connection with the more general problem of model choice.

Key words : nonparametric, prediction, regression, mean square error, kernel estimate, regressogram, predictogram.

A.M.S. 1980 subject classifications - Primary 62G05.

Acknowledgements : My sincere thanks are due to Professor Bosq, University of Lille, for the suggestion of this work on the predictogram, and to Professor Raoult, University of Rouen, for his helpful remarks and constructive criticisms.

1. INTRODUCTION

Let X be a random vector which is $\mathbb{R}^p$ valued, $p \in \mathbb{N}_*$, and Y be a real random variable (r.r.v.). We denote by

$$r(\cdot) = E(Y/X = \cdot) \tag{1.1}$$

the *regression* function of Y on X and consider the function

$$r_n(\cdot\ ;\ (X_i,Y_i),\ i = 1,\ \ldots,\ n)$$

which is an estimator of the regression function r from the sample (X_i,Y_i), $i = 1, \ldots, n$, of n independent random pairs which are distributed as the pair (X,Y).

Now, let $\{Z_i,\ i = 1,2,\ \ldots\}$ be a real stationary process. For all integers n, we want to *predict* Z_{n+1} from the sequence $\{Z_i,\ i = 1,\ \ldots,\ n\}$. If the process is known, the best (for a quadratic loss function) predictor is the conditional expectation

$$E(Z_{n+1}\ /\ [Z_1,\ \ldots,\ Z_n])$$

which is identical with

$$R(Z_{n-k+1},\ \ldots,\ Z_n) = E(Z_{n+1}\ /\ [Z_{n-k+1},\ \ldots,\ Z_n]) \tag{1.2}$$

when the process is *markovian* of order k. Then, under that last assumption but when the process $(Z_n)_{\mathbb{N}}$ is not known, a "natural" prediction of Z_{n+1} from the sequence $\{Z_i,\ i = 1,\ \ldots,\ n\}$ is the "estimation" $R_n(Z_{n-k+1},\ \ldots,\ Z_n)$ of the r.r.v. $R(Z_{n-k+1},\ \ldots,\ Z_n)$ with

$$R_n(\cdot) = r_N(\cdot\ ;\ ([Z_{i-k+1},\ \ldots,\ Z_i],\ Z_{i+1}),\ i = k,\ \ldots,\ n-1) \tag{1.3}$$

where $N = n-k$ and r_n is the regression estimator (which is considered at the beginning of the present introduction) when $p = k$.

Most parametric or non parametric predictors are defined in such a way from an estimator of the regression function. Here we only consider some non parametric methods.

It is clear that the non parametric estimation of regression is an area of non parametric statistics which now is almost as classical as the non parametric estimation of density. Results on non parametric *regression estimation* are reviewed by Collomb (1981), who also gives a review of reviews on *density estimation*.

Here, we focus our attention on the mean square error (m.s.e.) of various non parametric estimators (n.p.e.) of the regression function (we will consider results which are more recent than results reviewed by Collomb, 1981) and of non parametric predictors R_n defined by (1.3) where r_n is a n.p.e. of regression.

Since r_n and R_n are both functionnal estimators, we can define the m.s.e. in three different senses

(i) the pointwise m.s.e.

$$q_n(x) = E(r_n(x) - r(x))^2 \ , \quad x \text{ fixed in } \mathbb{R}^p \ , \tag{1.4}$$

$$Q_n(z) = E(R_n(z) - R(z)) \quad , \quad z \text{ fixed in } \mathbb{R}^k \ ;$$

(ii) the mean integrated square error (M.I.S.E., introduced by Parzen (1962) in connection with density estimation)

$$q_n^\lambda = \int_{\mathbb{R}^p} E(r_n(x) - r(x))^2 \ \lambda\ (dx), \tag{1.5}$$

$$Q_n^\lambda = \int_{\mathbb{R}^k} E(R_n(z) - R(z))^2 \ \lambda\ (dz),$$

where λ is a positive measure;

(iii) the mean mean square error

$$q_n = E(r_n(X) - r(X))^2 \tag{1.6}$$

where X is independent of (X_i,Y_i), $i = 1, \ldots, n$ (this is obviously a special case of (1.5)) and

$$Q_n = E(R_n(Z_{n-k+1}, \ldots, Z_n) - R(Z_{n-k+1}, \ldots, Z_n))^2.$$

The n.p.e. r_n and R_n are considered here from the point of view of their m.s.e. for the three following reasons

(i) the investigation of the limit of the m.s.e. gives directly a first asymptotic property;

(ii) results on the evaluation or the majoration of the m.s.e. give information on the *rate of convergence* of these n.p.e. : comparisons are possible, between various regression estimators and between these estimators and prediction function estimators;

(iii) as for non parametric predictors, nearly all the available results concern the m.s.e., at the present stage of development of the subject.

The present paper is made up of three parts, which are self-contained and of not equal importance :

part I : results on the mean square error of *regression* estimators : we review the results concerning the limit, majoration or evaluation of the quantities defined by (1.4), (1.5) and (1.6);

part II : short review of the few works existing in non parametric prediction;

part III : the mean square error of the *predictogram* : for this very simple non parametric predictor, we give our own recent results (*necessary and sufficient condition for* L_2 *consistency* and evaluation of the *optimal rate of convergence*).

The reader who is only interested by the original part of the present contribution can refer directly to these results, by proceeding to part III, which is independent of the parts I and II and of the present introduction.

2. REVIEW OF RESULTS ON THE MEAN SQUARE ERROR IN NON PARAMETRIC REGRESSION

We complete the definition (1.1) by

$$v(x) = E\Big((Y - r(X))^2 \;/\; X = x\Big), \quad \forall\, x \in \mathbb{R}^p, \tag{2.1}$$

and denote by f the density of the law of X with respect to the Lebesgue measure on $\mathbb{R}^p$.

2.1. The kernel method

Let K be a *kernel* of $\mathbb{R}^p$, that is to say a function in $L_1(\mathbb{R}^p)$, bounded and satisfying $|y|^p K(y) \xrightarrow[|y| \to \infty]{} 0$. We suppose that this kernel is positive, symmetrical and satisfies

$$\int_{\mathbb{R}^p} K(u)\, du = 1.$$

We consider the n.p.e

$$\mathring{r}_n(x) = \frac{\sum_{i=1}^{n} Y_i\, K\left(\frac{x - X_i}{h_n}\right)}{\sum_{i=1}^{n} K\left(\frac{x - X_i}{h_n}\right)}, \quad \left(\frac{0}{0} = 0\right) \quad \forall\, x \in \mathbb{R}^p, \tag{2.2}$$

where $\{h_n,\ n = 1, 2, \ldots,\}$ is a sequence of positive numbers satisfying

$$h_n \xrightarrow[n \to \infty]{} 0.$$

This estimator was proposed (independently and at the same time) by Watson (1964), who studied it by simulation, and Nadaraya (1964) who investigated its basic properties $(p = 1)$.

2.1.1. Limit and majoration of the pointwise m.s.e.

A consequence of Theorem 2 of Nadaraya (1964, p. 142) is that

$$E(\mathring{r}_n(x) - r(x))^2 \xrightarrow[n \to \infty]{} 0 \tag{2.3}$$

when Y is bounded and r and f are continuous at x, with $f(x) \neq 0$ and $n h_n^2 \xrightarrow[n \to \infty]{} \infty$.

Rosenblatt (1969) and Konakov (1973) give similar results. Noda (1976) obtained a majoration of the pointwise m.s.e. and used it in an investigation of the rate of convergence of $r_n(x)$, x fixed. All these papers only concern the case $p = 1$. Collomb (1976) proved that (2.3) holds if and only if

$$n h_n^p \xrightarrow[n \to \infty]{} \infty. \tag{2.4}$$

2.1.2. Evaluation of the pointwise m.s.e.

Collomb (1976 or 1977a) obtained the following expression of bias and variance of $\mathring{r}_n(x)$

. *bias* : if the second derivative $r''(x)$ of r and the first derivative $f'(x)$ of f at x exist, then

$$E\mathring{r}_n(x) - r(x) = h_n^2 \operatorname{tr}\Big(b(x)[K]\Big) + o\Big(h_n^2\Big) + o\Big(\frac{1}{nh_n^p}\Big) \tag{2.5}$$

where

$$b(x) = \Big[r''(x) + \Big(r'(x)^t f'(x) + f'(x)^t r'(x)\Big) / f(x)\Big] / 2 \tag{2.6}$$

and

$$[K] = \int_{\mathbb{R}^p} u\, {}^t u \, du \tag{2.7}$$

with "${}^t u$" denoting "transposed u" for all u in $\mathbb{R}^p$. In the case $p = 1$, this expression of bias becomes

$$E\mathring{r}_n(x) - r(x) = h_n^2 \left(\frac{r''(x)}{2} + r'(x) \frac{d \log f(x)}{dx}\right) \int z^2 K(z)\, dz + o\Big(h_n^2\Big) + o\Big(\frac{1}{nh_n^p}\Big) \tag{2.8}$$

. *variance* : if v and f are both continuous at x, than

$$E\Big(\mathring{r}_n(x) - E\mathring{r}_n(x)\Big)^2 = \frac{1}{nh_n^p} \frac{v(x)}{f(x)} \int K^2(u)\, du + o\Big(\frac{1}{nh_n^p}\Big) \tag{2.9}$$

The results (2.5) and (2.9) are coherent with the results of Rosenblatt (1969). A first consequence of these formulas is the following expression of the m.s.e.

$$E\Big(\mathring{r}_n(x) - r(x)\Big)^2 = \frac{1}{nh_n^p} \frac{v(x)}{f(x)} \int K^2(u)\, du + h_n^4 \operatorname{tr}^2 (b(x)[K]) + o\Big(h_n^4\Big) + o\Big(\frac{1}{nh_n^p}\Big) \tag{2.10}$$

2.1.3. Rate of convergence

The previous formula gives information about the pointwise optimal rate of convergence of the kernel method

$$\min_{h_n \in \mathbb{R}^+} E\Big(\mathring{r}_n(x) - r(x)\Big)^2 \underset{n \to \infty}{\sim} d(x) n^{-4/(p+4)} \tag{2.11}$$

where

$$d(x) = \frac{p+4}{4}\left(\frac{p}{4}\right)^{-p/(p+4)} \left(\frac{v(x)}{f(x)} \int K^2\right)^{4/(p+4)} \left(tr^2(b(x)[K])\right)^{p/(p+4)} \tag{2.12}$$

Elsewhere, for a suitable choice of λ and under assumptions (which especially concern the functions r'' and f') which authorize the integration of (2.10), the M.I.S.E. satisfies

$$\int E\left(\overset{\circ}{r}_n(x) - r(x)\right)^2 \lambda(dx) = \frac{A}{nh_n^p} + Bh_n^4 + o\left(h_n^4\right) + o\left(\frac{1}{nh_n^p}\right) \tag{2.13}$$

and therefore

$$\min_{h_n \in R_*^+} \int E\left(\overset{\circ}{r}_n(x) - r(u)\right)^2 \lambda(dx) \underset{n\to\infty}{\sim} C\, n^{-4/(p+4)} \tag{2.14}$$

where A, B and C are positive constants. We also note here that a consequence (see Collomb, 1977) of the formula (2.10) is that the optimal kernel is the kernel introduced by Epanechnikov (1969) in density estimation.

When the function B is Lipschitz, Spiegelman and Sacks (1980) proved that the "mean mean square error" satisfies

$$E\left(\overset{\circ}{r}_n(X) - r(X)\right)^2 = O\left(n^{-2/(2+p)}\right) \quad \text{if} \quad h_n = n^{-1/(2+p)}. \tag{2.15}$$

2.1.4. Universal consistency

Devroye and Wagner (1980) and Spiegelman and Sacks (1980, K being the indicator of a ball centered at 0 in $\mathbb{R}^p$) proved that the kernel method is "universally consistent" in the following sense (Stone, 1977) :

$$E|Y|^q < \infty \Rightarrow E\,|\overset{\circ}{r}_n(X) - r(X)|^q \xrightarrow[n\to\infty]{} 0, \ \forall\, q \leq 1 \tag{2.16}$$

when $(h_n)_{\mathbb{N}}$ satisfies (2.4).

2.2. The k_n-nearest neighbour methods

The k_n-nearest neighbour (k-NN) estimator of the regression is defined by

$$\tilde{r}_n(x) = \frac{1}{k_n} \sum_{j \in J_n(x)} Y_j ;\ \forall\, x \in \mathbb{R}^p \tag{2.17}$$

where

$$J_n(x) = \{i : X_i \text{ is one of the } k_n \text{ observations nearest to } x\}$$

with

$$\frac{k_n}{n} \xrightarrow[n\to\infty]{} 0 \quad \text{and} \quad k_n \xrightarrow[n\to\infty]{} \infty. \tag{2.18}$$

This definition can be extended in two directions :

. a first extension, proposed by Royall (1966), is the estimator

$$\tilde{r}_n^{(1)}(x) = \sum_{i=1}^{n} Y_i\, W_{ni}(x) \tag{2.19}$$

with

$$\sum_{i=1}^{n} W_{ni}(x) = 1,\ W_{ni}(x) \geq 0, \qquad \forall\, i = 1, \ldots, n,$$

and $W_{ni}(x) = v_{nj}$ where j is the order of $|x - X_j|$ inside the set $\{|x - X_j|, j = 1, \ldots, n\}$ and the sequence $\{v_{nj}, j = 1, \ldots, n|$ satisfies for all integer n

$$v_{n1} \geq v_{n2} \geq \ldots \geq v_{nn}$$

$$\max_{j=1,\ldots,n} v_{nj} \xrightarrow[n \to \infty]{} 0, \qquad \sum_{j=k_{n+1}}^{n} v_{nj} \xrightarrow[n \to \infty]{} 0;$$

. a second extension comes from the utilization of a kernel, which is introduced at the beginning of paragraph 2.1. This new n.p.e. is defined by

$$\tilde{r}_n^{(2)}(x) = \sum_{i=1}^{n} Y_i\, K\Big((x - X_i) / H_n\Big) \Big/ \sum_{i=1}^{n} K\Big((x - X_i) / H_n\Big) \tag{2.20}$$

where H_n is the Eucledian distance between x and its k_n^{th} nearest neighbour inside the set $\{X_i, i = 1, \ldots, n\}$. This generalization of estimators $\mathring{r}_n$ and $\tilde{r}_n$ was proposed by Collomb (1979 a,b) and Mack (1981).

The introduction of results on the m.s.e. of the estimators $\tilde{r}_n^{(1)}$ and $\tilde{r}_n^{(2)}$ is similar to the introduction of results on the kernel estimator r_n.

2.2.1. Limit of the pointwise m.s.e.

Devroye (1981 a for a sufficient condition, 1981 b for a necessary and sufficient condition) obtained the property

$$E\Big(\tilde{r}_n^{(1)}(x) - r(x)\Big)^2 \xrightarrow[n \to \infty]{} 0 \tag{2.21}$$

for x fixed, and Collomb (1979 a,b) proved that

$$E\Big(\tilde{r}_n^{(2)}(x) - r(x)\Big) \xrightarrow[n \to \infty]{} 0 \tag{2.22}$$

when $|Y|$ is bounded and f is continuous at x, $f(x) \neq 0$.

2.2.2. Evaluation of the pointwise m.s.e.

Mack (1981) obtained the following expressions of the bias and variance of $\tilde{r}_n^{(2)}(x)$:

. *bias :* if the functions r and f are both continuously differentiable up to the second order in a neighbourhood of x, with $f(x) \neq 0$, then

$$E\left(\tilde{r}_n^{(2)}(x) - r(x)\right) = \left(\frac{k_n}{n}\right)^{2/p} \frac{tr(b(x)[K]}{(cf(x))^{2/p}} + o\left(\left(\frac{k_n}{n}\right)^2\right) + 0\left(\frac{1}{k_n}\right) \tag{2.23}$$

where the matrix $b(x)$ and $[K]$ are defined by (2.6) and (2.7), and c is the volume of the unit ball in $\mathbb{R}^p$, that is to say

$$c = \pi^{p/2} / \Gamma((p+2)/2). \tag{2.24}$$

. *variance* : if v and f are continuously differentiable in a neighbourhood of x, then

$$E\left(\tilde{r}_n^{(2)}(x) - E\tilde{r}_n^{(2)}(x)\right)^2 = \frac{cv(x)}{k_n} \int K^2(u)\, du + o\left(\frac{1}{k_n}\right). \tag{2.25}$$

A first consequence of the formulas (2.23) and (2.25) is the following expression of the m.s.e.

$$E\left(\tilde{r}_n^{(2)} - r(x)\right)^2 = \frac{cv(x)}{k_n} \int K^2(u)\, du + \left(\frac{k_n}{n}\right)^{4/p} \frac{tr^2 b(x)[K]}{(cf(x))^{4/p}} + o\left(\frac{k_n}{n}\right)^{4/p} + o\left(\frac{1}{k_n}\right) \tag{2.26}$$

2.2.3. Rate of convergence

The previous formula gives information about the pointwise optimal rate of convergence of the k_n-NN method

$$\min_{k_n \in \mathbb{N}} E\left(\tilde{r}_n^{(2)}(x) - r(x)\right)^2 \sim d(x)\, n^{-4/(p+4)} \tag{2.27}$$

where $d(x)$ is the constant defined by (2.12). Elsewhere, we can see that for a suitable choice of λ and assumptions which allow the integration of (2.26), the M.I.S.E. of the k-NN kernel estimator satisfies

$$\int E\left(\tilde{r}_n^{2}(x) - r(x)\right)^2 \lambda(dx) = \frac{\tilde{A}}{k_n} + \tilde{B}\left(\frac{k_n}{n}\right)^{4/p} + o\left(\frac{k_n}{n}\right)^{4/p} + o\left(\frac{1}{k_n}\right) \tag{2.28}$$

and therefore

$$\min_{k_n \in \mathbb{N}} \int E\left(\tilde{r}_n^{(2)}(x) - r(x)\right)^2 \lambda(dx) \underset{n\to\infty}{\sim} \tilde{C}\, n^{-4/(p+4)} \tag{2.29}$$

where $\tilde{A}$, $\tilde{B}$ and $\tilde{C}$ are positive constants.

2.2.4. Universal consistency

Stone (1977) proved that the condition (2.18) is a necessary and sufficient condition for the "universal consistency" of the estimator $\tilde{r}_n^{(1)}$, that is to say

$$\text{"}E|Y|^q < \infty \Rightarrow E\left(\tilde{r}_n^{(1)}(X) - r(X)\right)^2 \xrightarrow[n\to\infty]{} 0\text{"}. \tag{2.30}$$

Remarks on the kernel method ($\overset{\circ}{r}_n$) and the k-NN method ($\tilde{r}_n$)

A reading of the previous results shows that these two n.p.e. of regression satisfy the same asymptotic properties.

The zero-limit of the m.s.e., that is to say the results (2.3), (2.21) and (2.22) or (2.16) and (2.30), is obtained under assumptions (2.4) and (2.18) which are very similar. The sequences

$$h_n \quad \text{and} \quad \left(\frac{k_n}{ncf(x)}\right)^{1/p}$$

have an identical part inside the expression of the bias-formulas (2.5) and (2.23)-, the variance-formulas (2.9) and (2.25) - and the m.s.e. -formulas (2.9) and (2.26) - at a fixed point. The minimal pointwise m.s.e. - left member in formulas (2.11) and (2.27) - are asymptotically identical. Lastly, we note that the rate of convergence are identical, when the pointwise m.s.e. and the M.I.S.E.-formulas (2.14) and (2.29) - are observed.

The mathematical results on bias and variance can be examined from an *intuitive* perspective. For instance, let us consider formulas (2.6) and (2.23) on the bias of $\overset{\circ}{r}_n(x)$ and $\tilde{r}^2_n(x)$ when $p = 1$ and

$$K(\cdot) = 1_{\{\cdot\in[-0.5,0.5]\}} .$$

If we fix h_n and suppose that the law of (X,Y) is such that the regression r is concave at x and the density f is constant on a neighbourhood of x, these formulas show that the bias is negative. We can "see" on the Figure 2-1 which illustrates this case that the estimated value $\overset{\circ}{r}_n(x)$ or $\tilde{r}_n(x)$ will be smaller than the actual value $r(x)$. It is also intuitively obvious that the more important the concavity or the convexity is, the more important the absolute value of the error will be.

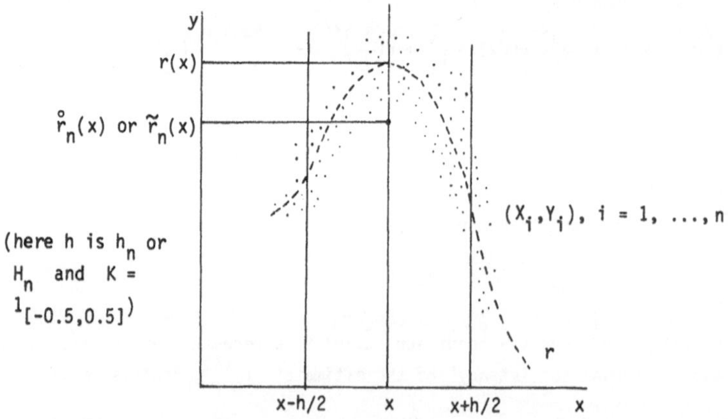

Figure 2.1 : Bias of the kernel estimator or of the k-NN estimator

Other similar intuitive considerations are stated by Collomb (1978), who also considers the following estimator.

2.4. The regressogram

This n.p.e. of regression was proposed by Tukey (1961) and first investigated by Bosq (1970). We consider just below its definition when X is a r.r.v. which is valued in $[0,1[$.

Let us divide the interval $[0,1]$ into equal and disjoint intervals of length h_n : the *regressogram* $\hat{r}_n$ is the function which, in each interval, is constant and equal to the average of the Y_i, $i = 1, \ldots, n$ such that X_i belongs to this interval (equal to zero if this interval does not contain an X_i).

This very simple definition can be easily generalized to the case $p > 1$ by

$$\begin{aligned} \hat{r}_n(x) &= \sum_{i=1}^{n} Y_i\, K_n(x,X_i) \Big/ \sum_{i=1}^{n} K_n(x,X_i) \quad \text{if} \quad \sum_{i=1}^{n} K_n(x,X_i) \neq 0 \\ &= 0 \qquad\qquad \text{otherwise}, \quad \forall\, x \in \mathbb{R}^p, \end{aligned} \tag{2.31}$$

where

$$K_n(x,u) = 1_{\left\{u \in \prod_{j=1}^{p} \left[h_n \operatorname{int}(x_j/h_n) \,/\, h_n(\operatorname{int}(x_j/h_n)+1)\right[\right\}}$$

We suppose that $(h_n)_{\mathbb{N}}$ satisfies

$$h_n \xrightarrow[n\to\infty]{} 0 \quad \text{and} \quad nh_n^p \xrightarrow[n\to\infty]{} \infty. \tag{2.32}$$

Collomb (1978) obtained the following expressions of bias and variance of $\hat{r}_n(x)$, x fixed in $\mathbb{R}^p$ and such that f is continuous and positive at x :

. *bias* : if the first derivative of r at x exists, then

$$E\hat{r}_n(x) - r(x) = h_n \sum_{j=1}^{p} \frac{\partial r(x)}{\partial x_j} \left(\left[\operatorname{int}\left(\frac{x_j}{h_n}\right) + \frac{1}{2}\right]h_n - x_j\right) + o(h_n) + o\left(\frac{1}{nh_n^p}\right) \tag{2.33}$$

. *variance* : if v is continuous at x, then

$$E\left(\hat{r}_n(x) - E\hat{r}_n(x)\right)^2 = \frac{1}{nh_n^p}\frac{v(x)}{f(x)} + o\left(\frac{1}{nh_n^p}\right) \tag{2.34}$$

The two previous results involve the following evaluation of the pointwise m.s.e. of the regressogram

$$\begin{aligned} E\left(\hat{r}_n(x) - r(x)\right)^2 &= \frac{1}{nh_n^p}\frac{v(x)}{f(x)} + h_n^2\left(\sum_{j=1}^{p} \frac{\partial r(x)}{\partial x_j}\left[\left(\operatorname{int}\left(\frac{x_j}{h_n}\right) + \frac{1}{2}\right) h_n - x_j\right]\right)^2 \\ &\quad + o(h_n) + o\left(\frac{1}{nh_n^p}\right) \end{aligned} \tag{2.35}$$

Therefore, for a suitable choice of λ and assumption which allow the integration of (2.35), the M.I.S.E. of the regressogram satisfies

$$\int E\left(\hat{r}_n(x) - r(x)\right)^2 \lambda(dx) \leq \frac{\hat{A}}{nh_n^p} + \hat{B}h_n^2 + o(h_n^2) + o\left(\frac{1}{nh_n^p}\right) \qquad (2.36)$$

which implies

$$\min_{h_n \in \mathbb{R}^+} \int E\left(\hat{r}_n(x) - r(x)\right)^2 \lambda(dx) \leq \hat{C}\, n^{-2/(p+2)} \qquad (2.37)$$

where $\hat{A}$, $\hat{B}$ and $\hat{C}$ are positive constants. We note that (2.36) and (2.37) remain valid only when r is Lipschitz.

Remarks

The previous results on the regressogram and the similar results on the kernel estimator and the k-NN estimator show that the regressogram is less precise but needs less restrictive assumptions on the law of (X,Y). In addition its utilization which is similar to the utilization of the *histogram* (n.p.e. of density) is very simple

2.5. Other methods

These other estimators are those of Priestley and Chao (1972), estimators defined with *orthogonal* functions or *spline* functions (Schlee, 1979, obtained an evaluation of the m.s.e. of spline estimators of density and regression), various *sequential* estimators and the regressogram with "random intervals" defined by *order statistics*. The papers on these estimators are reviewed by Collomb (1981).

2.6. General remarks

In connection with the investigation of the bias of these n.p.e. of regression, we mention the paper of Bickel and Lehman (1969) which implies (see Bosq, 1970) the *nonexistence of unbiased estimator* of the regression r (that is to say $\nexists\ r_n = r_n([X_i,Y_i], i = 1, \ldots, n) : \forall x \in \mathbb{R}^p,\ Er_n(x) = r(x))$ under the hypothesis which were introduced above on the law of the couple (X,Y)).

Lastly, let us remark that it can be verified that these results on the pointwise or integrated m.s.e. are coherent (see also the remark 6 of the paragraph 4 of the present paper) with general minimax type results on the optimal rate of convergence in function estimation (see e.g. Farrel, 1972, Meyer, 1977, Bretagnolle and Huber, 1979, Stone, 1980, or Birgé, 1980).

3. REVIEW OF PAPERS IN NON PARAMETRIC PREDICTION

The first paper in that field seems to be the article of Watson (1964, p. 369-370) who used the *kernel method* in a problem of meteorological prediction, with $k = 1$. The kernel predictor ρ_n is defined by (1.3) from (2.2), that is to say

$$\rho_n(z) = \sum_{i=1}^{n-1} Z_{i+1} K\left((z-Z_i) / h_n\right) / \left(\sum_{i=1}^{n-1} K((z-Z_i) / h_n)\right), \ \forall z \in \mathbb{R}, \tag{3.1}$$

where K is a kernel of $\mathbb{R}$ (for a definition, see the beginning of paragraph 2.1) and $\{h_n, n = 1,2, \ldots,\}$ is a sequence of positive members satisfying

$$h_n \xrightarrow[n \to \infty]{} 0.$$

Recently, Bosq (1981) and Doukhan (1981 a,b) completed the empirical investigation of Watson (1964) with various simulations.

The first mathematical study of this non parametric predictor is the paper of Roussas (1969) who proved the pointwise convergence in probability when (case $p = 1$)

$$nh_n \xrightarrow[n \to \infty]{} \infty.$$

Therefore, if we suppose that Z_1 is bounded, the pointwise m.s.e. satisfies

$$E\left(\rho_n(z) - R(z)\right)^2 \xrightarrow[n \to \infty]{} 0, \quad z \text{ fixed in } \mathbb{R}.$$

In the case of a process $(Z_n)_{\mathbb{N}}$ which is $\mathbb{R}^p$ valued, $p \geq 1$, but which is of the form

$$Z_{n+1} = R(Z_n) + \varepsilon_n$$

where $(\varepsilon_n)_{\mathbb{N}}$ is a sequence of i.i.d. random variables and R is continuously differentiable, Doukhan and Ghindès (1980) obtained the following majoration of the M.I.S.E.

$$\int E\left|\check{\rho}_n(z) - R(z)\right|^2 dz \leq An^{-2/(2+p)} \quad \text{if} \quad h_n = Bn^{-1/(2+p)} \tag{3.2}$$

where A and B are positive constants and $\check{\rho}_n$ is an estimator similar to (but not identical with) the estimator defined by (3.1).

Such an approach of the prediction problem is not very satisfying. Indeed, the "natural" problem is not the estimation of R for z fixed or on a subset of $\mathbb{R}^p$, but the natural problem is the "estimation" of $R(Z_n)$, Z_n being a random variable which by nature depends on Z_i, $i = 1, \ldots, n-1$. Bosq (1979) investigated the quantity

$$\int E|\hat{\rho}_n (Z_n) - R(Z_n)|^2$$

for a class of n.p.e. $\hat{\rho}_n$ which contains the kernel predictor ρ_n and the predictogram. However, Bosq (1979) supposed that the law of Z_1 is known. This assumption is rather restrictive, because of the non parametric character of the problem.

This twofold criticism is the fundation of our following work on the predictogram.

4. THE MEAN SQUARE ERROR OF THE PREDICTOGRAM

Preliminary

The present paragraph is self contained. We consider a problem which is more general than that considered in the introduction ($p = s = 1$, g = "identity", X_i denoted Z_i) in order to simplify the presentation of non parametric predictors from n.p.e. of regression.

The results mentioned below are proved by Collomb (1980).

4.1. Introduction

Let $(X_n)_{\mathbb{N}}$ be a *strictly stationary process* which is valued in a non empty part C of $\mathbb{R}^p$. We denote by g a real measurable function defined on $\mathbb{R}^p$ and by s and k two positive integers.

We consider the problem of the prediction of $g(X_{n+s})$ from the sequence X_i, $i = 1, \ldots, n$ and investigate the following non parametric method (see Figure 4.1).

Let P_n be a partition of C^k. We predict $g(X_{n+s})$ from X_i, $i=1, \ldots, n>k+s$ by the average of the $g(X_{i+s})$, $i = k, \ldots, n-s$ such that the vectors $(X_{i-k+1}, \ldots, X_i)$ and $(X_{n-k+1}, \ldots, X_n)$ belong to the same element of P_n, by 0 if there is no i for which this event is true.

This definition is similar to the definition of the *histogram* (n.p.e. of density), the *periodogram* (n.p.e. of spectral density) or the *regressogram* (n.p.e. of the regression, see 2.4). Therefore, we give the name of *predictogram* to the previous non parametric predictor. The predictogram can also be defined by the formula

$$\begin{aligned} R_n(X_{n-k+1}, \ldots, X_n) &= \frac{1}{|J|} \sum_{i=k}^{n-s} g(X_{i+s})\, 1_{\{(X_{i-k+1}, \ldots, X_i) \in J\}} \quad \text{if } |J| \neq 0, \\ &= 0 \qquad \text{otherwise;} \end{aligned} \tag{4.1}$$

with $J \in P_n : (X_{n-k+1}, \ldots, X_n) \in J$ and $|J| = \sum_{i=k}^{n-s} 1_{\{(X_{i-k+1}, \ldots, X_i) \in J\}}$.

When the process $(X_n)_{\mathbb{N}}$ is markovian of order k,

$$R(X_{n-k+1}, \ldots, X_n) = E(g(X_{n+s}) \;/\; [X_{n-k+1}, \ldots, X_n])$$

is the best *probabilist predictor* (Bosq, 1979) of $g(X_{n+s})$ if the process is known. When the process is unknown, the predictogram (4.1) is a *statistical predictor* (Bosq, 1979) which is considered here as a n.p.e. of this probabilist predictor. However, this Markov property is not useful in our investigation of the m.s.e. (called "mean mean square error" in the introduction)

$$Q_n = E\Big(R_n(X_{n-k+1}, \ldots, X_n) - R(X_{n-k+1}, \ldots, X_n)\Big)^2 \qquad (4.2)$$

Our assumptions on the process $(X_n)_{\mathbb{N}}$ will only be very few restrictive hypothesis of regularity.

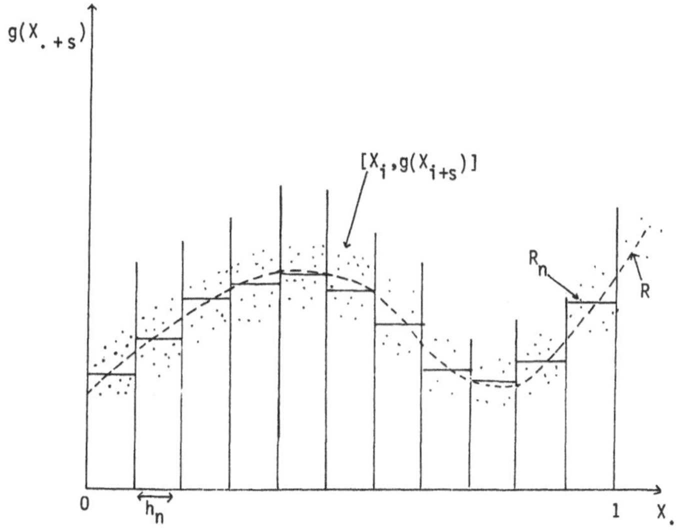

(here $C = [0,1[$ and P_n is the division of $[0,1[$ into equal and disjoint intervals of length h_n)

Figure 4.1 : An example of predictogram, in the case $p = k = 1$

4.2. Necessary and sufficient condition for L_2 consistency

We suppose that the process $(X_n)_{\mathbb{N}}$ takes values in

$$C = [0,1]^p$$

and that the function g is such that

$$|g(y)| \leq G, \ \forall\, y \in [0,1]^p \qquad (4.3)$$

where G is a positive constant.

For all partition P_n of $[0,1]^{kp}$, we denote

$$h_n = \inf\{h \in \mathbb{R}^+ : \forall B \in P_n, \exists u \in [0,1]^{kp} : B \subset [u-h, u+h]^{kp}\} \tag{4.4}$$

and suppose that

$$\exists d > 0 : \forall n \in \mathbb{N}, \mu(A) \geq d h_n^{kp}, \forall A \in P_n, \tag{4.5}$$

where μ is the Lebesque measure on $[0,1]^{kp}$: if each element of P_n is a cube of slide length h_n, then $d = 1$. We assume that

$$h_n \xrightarrow[n \to \infty]{} 0.$$

Lastly, *let $\mathcal{D}$ be the set of processes $(X_n)_{\mathbb{N}}$ such that*

$$\exists\, m,M,m > 0, 0 < M < \infty : m\mu(B) \leq P[(X_1, \ldots, X_k) \in B] \leq M\mu(B) \tag{4.6}$$

$$\forall B \in \mathcal{B}_{[0,1]^{kp}}$$

For each $(X_n)_{\mathbb{N}}$ in $\mathcal{D}$, we denote by ϕ_n a sequence of positive numbers for which this process $(X_n)_{\mathbb{N}}$ is ϕ-missing, in the following sense (Billingsley, 1968, p. 166) : for all integers $i > 0$ and $\ell > 0$, if A [resp. B] is a $\sigma(X_1, \ldots, X_i)$ [resp. $\sigma(X_{i+\ell}, X_{i+\ell+1}, \ldots)$] measurable set, then

$$|P(A \cap B) - P(A)\,P(B)| \leq \phi_\ell\, P(A).$$

First, we give a sufficient condition for the L_2-convergence of the predictogram.

THEOREM 1

If $(X_n)_{\mathbb{N}} \in \mathcal{D}$ and $(h_n)_{\mathbb{N}}$ satisfies

$$n h_n^{kp} \Big/ \sum_{\ell=0}^{n} \phi_\ell \xrightarrow[n \to \infty]{} \infty$$

then

$$Q_n \xrightarrow[n \to \infty]{} 0.$$

Remark 1

This result on the estimation of the *conditional expectation* $R(X_{n-k+1}, \ldots, X_n)$ is similar to the result on the estimation of the *expectation* of a bounded r.r.v. X_1 from a sequence $\{X_i, i = 1, \ldots, m\}$ which is stationary and ϕ-mixing, that is to say (Billingsley, 1968, formulas 20.32 and 20.29, p. 172)

$$\text{"}m \Big/ \sum_{\ell=0}^{m} \phi_\ell \xrightarrow[m \to \infty]{} \infty \Rightarrow \frac{1}{m} \sum_{i=1}^{m} X_i \xrightarrow[m \to \infty]{L_2} EX_1\text{"}$$

In prediction the term nh_n^{kp} takes a part which is similar to the one of the sample size m.

Further, we give a *necessary and sufficient* condition (on P_n) for the L_2-convergence of the predictogram. Let $\mathcal{D}_1$ be the set of processes in $\mathcal{D}$ such that

$$\sum_{\ell=0}^{\infty} \varphi_\ell \leq L < \infty \tag{4.7}$$

and suppose that there exists at least one process $(X_n)_{\mathbb{N}}$ in $\mathcal{D}_1$ such that $g \circ X_1$ is not a.s. constant.

THEOREM 2

When n *tends to infinity, the m.s.e.* Q_n *converges to* 0 *for every process* $(X_n)_{\mathbb{N}}$ *in* $\mathcal{D}_1$ *if and only if*

$$nh_n^{kp} \xrightarrow[n\to\infty]{} \infty$$

Remark 2

This result is similar to the results of Stone (1977), Devroye and Wagner (1980) or Spiegelman and Sacks (1980) on the property of "universal consistency" of some n.p.e. of regression : see paragraphs 2.2.4 and 2.3.4 of the present paper.

4.3. Majoration of the m.s.e. and rate of convergence

Now, we specify the result of the Theorem 1 by giving a *majoration* of the m.s.e. when the process $(X_n)_{\mathbb{N}}$ fulfils an additional condition which concerns the function R (Lipschitz condition)

$$\exists\, C > 0,\ C < \infty \text{ et } \gamma > 0 : |R(z) - R(z')| \leq C\|z-z'\|^\gamma,\ \forall\, z,z' \in [0,1]^{kp} \tag{4.8}$$

with $\|u_1, \ldots, u_{kp}\| = \max_{j=1,\ldots,kp} |u_j|$.

THEOREM 3

Let

$$t = k + s - 1,\quad N = n - t \quad\text{and}\quad \overline{\phi}_n = t + \sum_{i=1}^{N} \phi_i .$$

If the process $(X_n)_{\mathbb{N}}$ *is in* $\mathcal{D}$ *and satisfies the previous Lipschitz condition, then*

$$Q_n \leq M\left(\frac{4G}{md}\right)^2 \frac{1 + 8\overline{\phi}_n}{N\, h_n^{kp}} + 2c^2\, h_n^{2\gamma} + O\left(\frac{\overline{\phi}_n}{nh^{kp}}\right) \tag{4.9}$$

where the constants m, M, c, γ, d *and* G *are defined in* (4.6), (4.8), (4.5) and (4.3).

Remark 3

If we observe the above inequality, we can see that this mathematical result corroborates the following *intuitive* statements : for all fixed partition P_n (and therefore h_n) the precision of the statistical predictor is all the better as

. the dimension p of the vector X_1 or the number k is small

. the function R is "smooth" : the constant c is not too large and the order γ is not too small

. the law of $(X_1, \ldots, X_k)$ is "rather uniform" : the constants m and M are not too different

. the "dependence" between X_i and X_j, for all integers i and j, is small "in average" when $|i-j|$ is large : even if $\sum_{i=1}^{n} \phi_i$ tends to infinity as n increases, this quantity must remain small in comparison with n.

Remark 4

If moreover we have $\gamma = 1$ and $(X_n)_{\mathbb{N}}$ in $\mathcal{D}_1$, then the inequality (4.9) becomes

$$Q_n \leqslant \frac{a^2}{Nh_n^{kp}} + 2\,c^2\,h_n^2 + O\left(\frac{1}{nh_n^{kp}}\right)^2 \quad \text{with} \quad a^2 = \left(\frac{4G}{md}\right)^2 M[1+8(t + \sum_{\ell=0}^{\infty} \varphi_\ell)]$$

When $k = 1$, this majoration of the m.s.e. of the predictogram is similar to the majoration of the m.s.e. of the regressogram : see the formula (2.36) of the present paper.

The inequality (4.9) entails an information on the *optimal rate of convergence* of the predictogram. This information is given by a majoration of the minimum of the function $P_n \multimap Q_n(P_n)$ where $Q_n(P_n)$ is the m.s.e. defined by (4.2) for the predictogram which is associated with the partition P_n.

THEOREM 4

Under the assumptions of Theorem 3, we have

$$\inf_{P_n \in \Pi} Q_n(P_n) \leqslant A\left(\frac{\overline{\phi}_n}{n}\right)^{2\gamma/(2\gamma + kp)}$$

where Π *is the set of all the partitions of* $[0,1]^{kp}$ *which satisfy* (4.4) *and* (4.5), *and* A *is a positive constant.*

Remark 5

If moreover we have $\gamma = 1$ and $(X_n)_{\mathbb{N}}$ in $\mathcal{D}_1$, then

$$\inf_{P_n \in \Pi} Q_n(P_n) \leqslant An^{-2/(kp+2)}, \quad 0 < A < \infty.$$

The optimal rate of convergence which is given by the right hand side of the above inequality is identical with the optimal rate of convergence of the histogram or the regressogram (see formula 2.37) when the M.I.S.E. is considered.

Remark 6

This result on a n.p.e. of the function R defined on $\mathbb{R}^p$ and Lipschitz of order 1 is similar to the results of Parzen (1962), Collomb (1977, see the paragraph 2.1.3. of the present paper), Spiegelman and Sacks (1980, see formula 2.15 in the present paper) or Doukhan and Ghindès (1980, see the formula (3.2) of the present paper) on the rate of convergence of n.p.e. of a density, a regression or a prediction function, defined on $\mathbb{R}^p$ and differentiable up to the order q, $q \in \mathbb{N}_*$: all these n.p.e. reach the rates of convergence which are given by Bretagnolle and Huber (1979) and Birgé (1980) in connection with the investigation of minimax risks in non parametric estimation of density (see also the paragraph 2.5 of the present paper).

4.4. General remarks

These remarks concern the assumptions on the process $(X_n)_{\mathbb{N}}$, the possible applications or extensions of our results and lastly the utilization of the predictogram.

4.4.1. Markov property and ϕ-mixing condition

In the introduction, we did not suppose that the process was markovian. However we noted that a natural utilization of the predictogram is for the prediction in a Markov process. Therefore a natural question is : for k = 1, when does a Markov process satisfy the conditions (4.6) and (4.7) ? The following answer to this question comes from Doob (1953, p. 197, case (b) : this result is less general but more convenient than the result (-, 1953, p. 221) which uses Doeblin's condition)

> Let f be the transition density probability (that is to say $f(\cdot,u) = dP(X_{n+1} / X_n = u) / d\psi$, $\forall u \in C$) and for all integer j let $f^{(j)}$ be the j-step transition density defined by
>
> $$f^{(i)}(v,u) = \int_C f^{(i-1)}(v,z)\, f(z,u)\, \psi(dz),\ i = 1, \ldots, j,\ f^{(1)} = f.$$
>
> If
>
> $$\exists\, \nu \in \mathbb{N}, \delta > 0 : f^{(\nu)}(v,u) > \delta,\ \forall u,v \in C, \tag{4.9}$$
>
> then $(X_n)_{\mathbb{N}}$ is ϕ-mixing, with $\phi_\ell = 2(1-\delta)^{\ell/\nu-1}$, $\forall \ell \in \mathbb{N}$, and the stationary probability P satisfies $P(B) \geq \delta\mu(B)$, $\forall B \in \mathcal{B}_C$.

Consequently, the hypothesis (4.9) implies (4.7) with

$$L = 2(1-\delta)^{-1}\,(1-(1-\delta)^{1/\nu})^{-1}$$

and, in (4.6)

$$m = \delta.$$

4.4.2. Applications and extensions of results on the predictogram

A n.p.e. of the *transition probability distribution function* (t.p.d.f.)

$$F(u;[X_{n-k+1}, \dots, X_n]) = P(X_{n+s} \leq u/[X_{n-k+1}, \dots, X_n]), \forall u \in C$$

is

$$F_n(u;[X_{n-k+1}, \dots, X_n]) = \frac{1}{|J|} \sum_{i=k}^{n-s} 1_{\{X_{i+s} \leq u\}} 1_{\{[X_{i-k+1}, \dots, X_i] \in J\}}, \; 0/0 = 0 \qquad \forall u \in C$$

where J and $|J|$ are defined in (4.1). Collomb (1980, § 4) states some asymptotic properties of this new n.p.e. : refer to Theorems 1, ..., 4 of the present paper and consider the special case

$$g(z) = 1_{\{z \leq u\}}, \forall z \in [0,1]^p.$$

These results on the t.p.d.f. can be used in the investigation of n.p.e. of *conditional quantities* or of *conditional densities* which can be defined with F_n.

In addition, n.p.e. of *conditional moments* (when $p = 1$) are defined by the special case $g(z) = z^\beta$, $\beta \in \mathbb{N}$, directly from (3.1).

Lastly, the techniques of proof which are used in the investigation of the predictogram can be used in the investigation of the *histogram*

$$\pi_n(z) = \frac{1}{nh_n^{kp}} \sum_{i=k}^{n-s} 1_{\{[X_{i-k+1}, \dots, X_i] \in J\}}, \; J \in P_n : z \in J, \forall z \in C,$$

considered as a n.p.e. of the density function of the distribution of $[X_1, \dots, X_k]$.

4.4.3. Utilization of the predictogram

The definition and the utilization of the predictogram are similar to the definition and the utilization of the regressogram (see paragraph 2.3. in the present paper) and the histogram. They present the same characteristics :

. simplicity of use, especially if this method is compared with some classical parametrical methods (Box and Jenkins) which are more popular or with methods of prediction which are defined by n.p.e. of the spectral density (for instance see Parzen, 1957);

. problem of the choice of the partition P_n : here we note that the freedom in the choice of this partition authorizes the use of any "a priori" knowledge about the law of the process $(X_n)_{\mathbb{N}}$;

. difficulty of visualization if $kp > 1$;

. absence of precision, which is inherent to non parametrical methods, in regard to parametrical methods.

However it is clear that the practical interest of the histogram is its utilization in a first approach of the determination of a probability law. Likewise, the most important interest of the predictogram is its utilization in a first approach of a prediction problem for the determination of a parametric model.

5. APPENDIX. UTILIZATION OF THE PREDICTOGRAM : AN HEURISTIC METHOD FOR THE CHOICE OF THE PARTITION

The present appendix can be considered as a contribution to the field of data analysis. The method which is presented here is an extension of a *cross-validation method* which is given by Collomb (1978, annexe A, "Estimateur adaptatif") for general n.p.e. of regression.

We consider the following definition, in the case $C = \mathbb{R}^p$, of the predictogram

$$R_n^{(h)}(u) = \frac{\sum_{i=k}^{n-s} g(X_{i+s})\, K^{(h)}(u, [X_{i-k+1}, \ldots, X_i])}{\sum_{i=k}^{n-s} K^{(h)}(u, [X_{i-k+1}, \ldots, X_i])} \quad \left(\frac{0}{0} = 0\right), \ \forall\, u \in \mathbb{R}^{kp}$$

with

$$K^{(h)}(u,v) = 1_{\left\{v \in \prod_{j=1}^{p} \left[h\, \mathrm{int}\left(\frac{u_j}{h}\right),\ h\left(\mathrm{int}\left(\frac{u_j}{h}\right) + 1\right)\right[\right\}}, \ \forall\, u,v \in \mathbb{R}^{kp}$$

where h is a positive number (u_j, $j = 1, \ldots, kp$ denotes the jth composant of the vector u in $\mathbb{R}^{kp}$).

An illustration of this definition in the case $p = k = 1$ is given by the Figure 4.1 where h_n is replaced by h.

Some asymptotic properties of this predictogram are stated in the previous paragraphs when h depends on n. However, it is clear that the most important practical problem is : if we consider a sequence $\{X_i,\ i = 1, \ldots, n\}$, with n fixed, how to choose the number h ?

We propose the following method

let

$$R_{n,j}^{(h)} = \frac{\sum_{\substack{i=k \\ i \neq j}}^{n-s} g(X_{i+s})\, K^{(h)}([X_{j-k+1}, \ldots, X_j], [X_{i-k+1}, \ldots, X_i])}{\sum_{\substack{i=k \\ i \neq j}}^{n-s} K^{(h)}([X_{j-k+1}, \ldots, X_j], [X_{i-k+1}, \ldots, X_i])}, \quad j = k, \ldots, n-s$$

(A) | *with* $0/0 = 0$, *and*

$$S_n(h) = \sum_{j=k}^{n-s} \left(g(X_{j+s}) - R_{n,j}^{(h)}\right)^2,$$

then choose an h^* *satisfying*

$$S_n(h^*) = \min_{h \in \mathbb{R}_*^+} S_n(h).$$

It is clear that most of the practicians make approximatively the choice of h, when they choose the number h so that on the same graphical display (if $k = p = 1$, see Figure 4.1) the curve R_n is "well centered" inside the set $\{(X_i, g(X_{i+s})),\ i = k, \dots, n-s\}$. The algorithm (A) is only a formalization of this intuitive approach.

The statistical predictor $R_n^{(h^*)}(\cdot)$, where h^* is the r.r.v. defined by (A), is a new n.p.e. of the prediction function. This n.p.e. seems to deserve to be studied from a mathematical point of view.

REFERENCES

Bickel, P.J., Lehman, E.L. (1969), "Unbiased Estimation in Convexe Families", *Ann. Math. Stat.* 40, 1523-1525.

Billingsley, P. (1968), "*Convergence of Probability Measures*", New-York, Wiley.

Birgé, L. (1980), "Approximation dans les espaces métriques et théorie de l'estimation; inégalités de Cramer-Chernoff et théorie asymptotique des tests", Thèse, Université de Paris VII.

Bosq, D. (1970), "Contribution à la théorie de l'estimation fonctionnelle", *Publications de l'Institut de Statistique de l'Université de Paris*, 19, fasc. 2 et 3.

Bosq, D. (1979), "Sur la prédiction non paramétrique de variables aléatoires et de mesures aléatoires", publication interne, U.E.R. de Maths, Lille.

Bosq, D. (1981, "Non Parametric Prediction for a Stationary Process", Communication lors de la rencontre franco belge de statisticiens, Louvain, 1981, to appear in *Lecture Notes*.

Bretagnolle, J., Huber, C. (1979), "Estimation des densités : risques minimaux", *Z. Wahrscheinlichkeitstheorie. Verw. Geb.* 47, 119-137.

Collomb, G. (1976), "Estimation non paramétrique de la régression par la méthode du noyau", Thèse, Université Paul Sabatier, Toulouse.

Collomb, G. (1977), "Quelques propriétés de la méthode du noyau pour l'estimation non paramétrique de la régression en un point fixé", *Comptes Rendus Acad. Sci. Paris* 285, Série 1, 289-292.

Collomb, G. (1978), "Estimation non paramétrique de la régression : régressogramme et méthode du noyau", *Publications du Laboratoire de Statistique et Probabilités de l'Université de Toulouse*, 07-78, 1-59.

Collomb, G. (1979a), "Estimation de la régression par la méthode des k points les plus proches : propriétés de convergence ponctuelle", *Comptes Rendus Acad. Sci. Paris*, 289, Série A, 245-247.

Collomb, G. (1979b), "Estimation de la régression par la méthode des k points les plus proches avec noyau", Communication lors des Journées S.M.F. de Rouen, Juin 1979, *Lec. Notes Math.*, 821, 159-175.

Collomb, G. (1980), "Prédiction non paramétrique : étude de l'erreur quadratique du prédictogramme", publication interne, Laboratoire de Statistique, Université Paul Sabatier, Toulouse.

Collomb, G. (1981), "Estimation non-paramétrique de la régression : Revue Bibliographique", *Int. Stat. Rev.*, 49, 75-93.

Devroye, L.P., Wagner, T.J. (1980), "Distribution-Free Consistency Results in Nonparametric Discrimination and Regression Function Estimation", *Ann. Stat.* 8, 231-239.

Devroye, L.P. (1981a), "On the Almost Everywhere Convergence of Nonparametric Regression Function Estimates", *Ann. Stat.* 9, 1310-1319.

Devroye, L.P. (1981b), "Necessary and Sufficient Conditions for the Pointwise Convergence of Nearest Neighbour Regression Function Estimates", Preprint.

Doob, J. (1953), "*Stochastic Processes*", New-York, Wiley.

Doukhan, P., Ghindès, M. (1980), "Estimations dans le processus '$X_{n+1} = f(X_n) + \varepsilon_n$'", *Comptes Rendus Acad. Sci. Paris*, Série A, 297, 61-64.

Doukhan, P. (1981a), "Simulations dans le processus autorégressif général d'ordre 1; cas unidimensionnel", prépublication, Université de Paris Sud, Département de Mathématiques.

Doukhan, P. (1981b), "Simulations in the General First Order Autogressive Process (Unidimensional Normal Case)", Rencontre franco belge de statisticiens, Louvain, 1981, to appear in *Lecture Notes*.

Epanechnikov, V.A. (1969), "Non Parametric Estimation of a Multivariate Density", *Theory Probab. Appl.* 14, 153-158.

Farrell, R.H. (1972), "On the Best Obtainable Asymptotic Rates of Convergence in Estimation of a Density Function at a Point", *Ann. Math. Stat.* 43, 170-180.

Konakov, V.D. (1972), "Asymptotic Properties of Some Functions of Nonparametric Estimates of a Density Function", *J. Multivariate Anal.* 5, 454-468.

Mack, Y.P. (1981), "Local Properties of k-NN Regression Estimates", to appear in *SIAM*.

Meyer, T.G. (1977), "Bounds for Estimation of Density Functions and their Derivatives", *Ann. Stat.* 5, 136-142.

Nadaraya, E.A. (1964), "On Estimating Regression", *Theory Probab. Appl.* 9, 141-142.

Noda, K. (1976), "Estimation of a Regression Function by the Parzen kernel-type Density Estimators", *Ann. Inst. Stat. Math.*28, 221-234.

Parzen, E. (1957), "On Consistent Estimates of the Spectum of a Stationary Time Series", *Ann. Math. Stat.* 28, 329-348.

Parzen, E. (1962), "On Estimation of a Probability Density and Mode", *Ann. Math. Stat.* 35, 1065-1076.

Priestley, M.B., Chao, M.T. (1972), "Non-Parametric Function Fitting", *J. R. Stat. Soc., Ser. B.* 34, 385-392.

Rosenblatt, M. (1969), "Conditional Probability Density and Regression Estimators", *Multivariate Analysis II,* 25-31, Academic Press, New York.

Roussas, G. (1969), "Nonparametric Estimation of the Transition Distribution Function of a Markov Process", *Ann. Math. Stat.* 40, 1386-1400.

Royall, R.M. (1966), "A Class of Non Parametric Estimators of Smooth Regresstion Function", Ph.D. Dissertation, Stanford University.

Schlee, W. (1979), "Non Parametric Estimation of Curves", *Serdica* 5, 186-203.

Spielgelman, G., Sacks, J. (1980), "Consistent Window Estimation in Nonparametric Regression Estimation", *Ann. Stat.* 8, 240-246.

Stone, C.J. (1977), "Consistent Nonparametric Regression", *Ann. Stat.* 5, 595-645, with the contribution of Bickel, P.J., Breiman, L. Brillinger, D.R., Brunk, H.D., Pierce, D.A., Chernoff, H., Cover, T.M., Cox, D.K., Eddy, W.F., Hampel, F., Olshem, R.A., Parzen, E., Rosenblatt, M., Sacks, J., Wahba, G.

Stone, C.J., (1980), "Optimal Rates of Convergence for Non Parametric Estimators", *Ann. Stat.* 8, 6, 1348-1360.

Tukey, J.W. (1961), "Curves as Parameters and Touch Estimation", *Proceedings of the 4th Berkeley Symposium on Mathematical Statistics and Probability,* 681-694.

Watson, G.S. (1964), "Smooth Regression Analysis", *Sankhya, Ser. A* 26, 359-372.